Studies in Computational Intelligence

Volume 757

Series editor

Janusz Kacprzyk, Polish Academy of Sciences, Warsaw, Poland
e-mail: kacprzyk@ibspan.waw.pl

The series "Studies in Computational Intelligence" (SCI) publishes new developments and advances in the various areas of computational intelligence—quickly and with a high quality. The intent is to cover the theory, applications, and design methods of computational intelligence, as embedded in the fields of engineering, computer science, physics and life sciences, as well as the methodologies behind them. The series contains monographs, lecture notes and edited volumes in computational intelligence spanning the areas of neural networks, connectionist systems, genetic algorithms, evolutionary computation, artificial intelligence, cellular automata, self-organizing systems, soft computing, fuzzy systems, and hybrid intelligent systems. Of particular value to both the contributors and the readership are the short publication timeframe and the world-wide distribution, which enable both wide and rapid dissemination of research output.

More information about this series at http://www.springer.com/series/7092

M Hadjiski · K T Atanassov
Editors

Intuitionistic Fuzziness and Other Intelligent Theories and Their Applications

 Springer

Editors
M Hadjiski
University of Chemical Technology
  and Metalurgy
Sofia
Bulgaria

K T Atanassov
Department of Bioinformatics
  and Mathematical Modelling, Institute
  of Biophysics and Biomedical
  Engineering
Bulgarian Academy of Sciences
Sofia
Bulgaria

ISSN 1860-949X   ISSN 1860-9503   (electronic)
Studies in Computational Intelligence
ISBN 978-3-030-07683-2   ISBN 978-3-319-78931-6   (eBook)
https://doi.org/10.1007/978-3-319-78931-6

Printed on acid-free paper

This Springer imprint is published by the registered company Springer International Publishing AG
part of Springer Nature
The registered company address is: Gewerbestrasse 11, 6330 Cham, Switzerland

Preface

Since the beginning of the new century, every even year, the IEEE Conference on Intelligent Systems had been organized. The first, second, and fourth editions (in 2002, 2004, and 2008, respectively) were organized in Varna, Bulgaria. The third and fifth one (in 2006 and 2010, respectively) took place in London, UK, and the sixth conference (in 2012) was in Sofia, Bulgaria. Then, the next, seventh conference (in 2014) was held in Warsaw, Poland, and the last, eight conference (in 2016)—again in Sofia, Bulgaria. The 2016 IEEE Conference on Intelligent Systems, IEEE IS'2016, was technically sponsored by the IEEE Systems, Man, and Cybernetics Society.

Preserving the tradition that had been accepted since the first edition of the IEEE Conference on Intelligent Systems, the members of the Program Committee of IEEE Intelligent Systems 2016 had again selected the best papers part of which had been included in the present volume.

Generally speaking, the papers selected are mainly related to more foundational and theoretical research results in the area of intelligent systems, fuzzy and intuitionistic fuzzy sets, generalized nets, and intercriteria analysis, to only mention the most representative fields.

Chapter "Intercriteria Analysis and Arithmetic Functions" by K T Atanassov, Vassia Atanassova, and Panagiotis Chountas is devoted to applying the method of intercriteria analysis over values of arithmetic functions. The intercriteria analysis is a new method for the comparison of objects, different from the correlation analyses (due to Pearson, Spearman, etc.). On the one hand, the results presented are an illustration of this new method and, on the other hand, they reveal some currently unknown properties of the case of arithmetic functions, suggesting some future research problems too.

In Chapter "Fuzzy Harmony Search Algorithm Using an Interval Type-2 Fuzzy Logic Applied to Benchmark Mathematical Functions" by Cinthia Peraza, Fevrier Valdez, and Oscar Castillo, a fuzzy harmony search algorithm based on an interval type-2 fuzzy logic system for dynamic parameter adaptation is described. It is successfully applied to various benchmark optimization problems. Numerical results show that the proposed algorithm can find better solutions than those obtained by using a type-1 fuzzy harmony search and other heuristic methods.

Chapter "Mixture Initialization Based on Prior Data Visual Analysis" by Evgenia Suzdaleva and Ivan Nagy is related to one of the key problems of the mixture estimation—the initialization. It discusses the initialization of the recursive mixture estimation for the case of normal components. Several initialization techniques based on a visual analysis of prior data are given. Validation experiments are presented.

In Chapter "Spatiotemporal Parameter Estimation of Thermal Treatment Process via Initial Condition Reconstruction Using Neural Networks" by M Hadjiski, Nencho Deliiski, and Aleksandra Grancharova, the design of control systems of periodical thermal treatment processes with distributed parameters modeled by partial differential equations is considered. The investigation is based on the first principle models of the internal and external heat exchange. Some aspects of industrial application of the proposed approaches are discussed.

A differential evolution algorithm, using fuzzy logic to make dynamic changes in the mutation and crossover parameters separately, is proposed in Chapter "Interval Type-2 Fuzzy Logic Dynamic Mutation and Crossover Parameter Adaptation in a Fuzzy Differential Evolution Method" by Patricia Ochoa, Oscar Castillo, and José Soria. A comparison of the fuzzy differential evolution algorithm using the type-1 fuzzy logic and the interval type-2 fuzzy logic is performed for a set of benchmark functions.

A method for evaluation of the student's knowledge obtained in the university e-learning courses is described in Chapter "Intuitionistic Fuzzy Evaluations for Analysis of a Student's Knowledge in University e-Learning Courses" by Evdokia Sotirova, Anthony Shannon, Taekyun Kim, Maciej Krawczak, Pedro Melo-Pinto, and Beloslav Riečan. It provides the possibility for the algorithmization of the process of formation of student course assessments.

Chapter "S-Logic with First and Second Imaginary States" by Vassil Sgurev is devoted to the introduction of a second imaginary j-logic. Insolvability in the i-logic is surmounted. For this purpose, constraints are ushered in between the variables of the classical r-logic and those of the i-logic and j-logic. This gives rise to the complex logics—s, s_1, s_2, respectively, in which the functioning of the r, i, and j-logics is being interpreted. The field of application of the logical structures being investigated is discussed.

The processes in a center of transfusion hematology—receiving blood, obtaining fresh frozen plasma, erythrocytes and thrombocytes, its testing for transmitted diseases (HIV, HBV, HCV, and Wass) and evaluation of a blood group and the Rh, and antibodies screening—are modeled by using the generalized nets in Chapter "Generalized Net Model of the Processes in a Center of Transfusion Haematology" by Nikolay Andreev, Evdokia Sotirova, Anthony Shannon, and K T Atanassov.

Chapter "Image to Sound Encryption Using a Self-Organizing Map Neural Network" by Todor Petkov and Sotir Sotirov contains the description of the process of encrypting image in a sound using an artificial neural network. In order to attain this, the process is divided into several steps where each of the steps is described by a generalized net.

Chapter "On Different Algorithms for InterCriteria Relations Calculation" by Olympia Roeva, Peter Vassilev, Nikolay Ikonomov, Maria Angelova, Jun Su, and Tania Pencheva describes the use of the recently proposed InterCriteria analysis approach for searching of existing or unknown correlations between multiple objects against multiple criteria. Altogether, five different algorithms for the calculation of intercriteria relations are examined to render the influence of the genetic algorithm parameters on the algorithm performance.

In Chapter "Defining Consonance Thresholds in InterCriteria Analysis: An Overview" by Lyubka Doukovska, Vassia Atanassova, Evdokia Sotirova, Ivelina Vardeva, and Irina Radeva, a detailed overview is given of the development of the approaches adopted in defining the consonance thresholds in the recently proposed method for decision support, the so-called InterCriteria analysis. The gradual progress in defining the thresholds of the membership and non-membership parts of the intuitionistic fuzzy pairs serves as estimations of the pairwise consonances.

Chapter "Design and Comparison of ECG Arrhythmias Classifiers Using Discrete Wavelet Transform, Neural Network and Principal Component Analysis" by Seyed Saleh Mohseni and Vahid Khorsand is devoted to an important aspect of the today's medical systems—the automatic classification of heartbeat. In the paper, two separate classifiers are designed and compared for heartbeat classification.

We would like to express our gratitude to all the authors for their interesting, novel, and inspiring contributions. It should also be noted that many of the authors are young scientists. Peer reviewers also deserve our deep appreciation because of their insightful and constructive remarks and suggestions that have considerably improved the contributions.

And last but not least, we wish to thank Dr. Tom Ditzinger, Dr. Leontina di Cecco, and Mr. Holger Schaepe for their dedication and help to implement and finish this large publication project on time maintaining the highest publication standards. It is important to note that a lot of the authors are young specialists.

Sofia, Bulgaria   M Hadjiski
                                                                                        K T Atanassov

Contents

Intercriteria Analysis and Arithmetic Functions

K T Atanassov, Vassia Atanassova and Panagiotis Chountas

Abstract The possibility to apply the intercriteria analysis over normalized data is discussed. An example, related to evaluation of six arithmetic functions, is given.

Keywords Arithmetic function · Intercriteria analysis · Intuitionistic fuzziness

1 Introduction

The concept of InterCriteria Analysis was introduced in [1, 2], on the basis of apparatus of the Index Matrices (IMs, see [1]) and of Intuitionistic Fuzzy Pairs (IFPs, see, e.g., [3]) and Sets (IFSs, see, e.g., [4]). The paper is a continuation of [2, 5, 6].

In Sects. 2 and 3, short notes on IFPs and of IMs are given, respectively. In Sect. 4, we discuss the possibility for the data to be processed by intercriteria analysis. In Sect. 5, an example, related to arithmetic functions is given.

K T Atanassov (✉) · V. Atanassova
Department of Bioinformatics and Mathematical Modelling, Institute of Biophysics and Biomedical Engineering, Bulgarian Academy of Sciences, Acad. G. Bonchev Str., Block 105, 1113 Sofia, Bulgaria
e-mail: krat@bas.bg

V. Atanassova
e-mail: vassia.atanassova@gmail.com

K T Atanassov
Prof. Asen Zlatarov University Burgas, 1 Prof. Yakimov Blvd, 8010 Burgas, Bulgaria

P. Chountas
Faculty of Science and Technology, Department of Computer Science, University of Westminster, 115 New Cavendish Street, London W1W 6UW, UK
e-mail: P.I.Chountas@westminster.ac.uk

© Springer International Publishing AG, part of Springer Nature 2019
M Hadjiski and K T Atanassov (eds.), *Intuitionistic Fuzziness and Other Intelligent Theories and Their Applications*, Studies in Computational Intelligence 757,
https://doi.org/10.1007/978-3-319-78931-6_1

2   Short Notes on Intuitionistic Fuzzy Pairs

The Intuitionistic Fuzzy Pair (IFP) is an object in the form $\langle a, b \rangle$, where $a, b \in [0, 1]$ and $a + b \leq 1$, that is used as an evaluation of some object or process and which components (a and b) are interpreted as degrees of membership and non-membership, or degrees of validity and non-validity, or degree of correctness and non-correctness, etc. One of the geometrical interpretations of the IFPs is shown on Fig. 1.

Let us have two IFPs $x = \langle a, b \rangle$ and $y = \langle c, d \rangle$. We define the relations

$$x < y \quad \text{iff} \quad a < c \text{ and } b > d$$
$$x > y \quad \text{iff} \quad a > c \text{ and } b < d$$
$$x \geq y \quad \text{iff} \quad a \geq c \text{ and } b \leq d$$
$$x \leq y \quad \text{iff} \quad a \leq c \text{ and } b \geq d$$
$$x = y \quad \text{iff} \quad a = c \text{ and } b = d$$

3   Short Remarks on Index Matrices

The concept of Index Matrix (IM) was discussed in a series of papers (see, e.g., [1]).

Let I be a fixed set of indices and $\mathcal{R}$ be the set of the real numbers. By IM with index sets K and L ($K, L \subset I$), we denote the object:

$$[K, L, \{a_{k_i, l_j}\}] \equiv \begin{array}{c|cccc} & l_1 & l_2 & \dots & l_n \\ \hline k_1 & a_{k_1, l_1} & a_{k_1, l_2} & \dots & a_{k_1, l_n} \\ k_2 & a_{k_2, l_1} & a_{k_2, l_2} & \dots & a_{k_2, l_n} \\ \vdots & \vdots & \vdots & \ddots & \vdots \\ k_m & a_{k_m, l_1} & a_{k_m, l_2} & \dots & a_{k_m, l_n} \end{array},$$

where $K = \{k_1, k_2, ..., k_m\}$, $L = \{l_1, l_2, ..., l_n\}$, for $1 \leq i \leq m$, and $1 \leq j \leq n$: $a_{k_i, l_j} \in \mathcal{R}$.

In [1], different operations, relations and operators are defined over IMs. For the needs of the present research, we will introduce the definitions of some of them.

When elements a_{k_i, l_j} are some variables, propositions or formulas, we obtain an extended IM with elements from the respective type. Then, we can define the evalu-

Fig. 1 Correspondence of the element of the intuitionistic fuzzy set with membership and non-membership values, respectively $\langle a, b \rangle$ to a point onto the intuitionistic fuzzy interpretational triangle

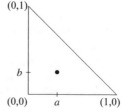

ation function V that juxtaposes to this IM a new one with elements—IFPs $\langle \mu, \nu \rangle$, where $\mu, \nu, \mu + \nu \in [0, 1]$. The new IM, called Intuitionistic Fuzzy IM (IFIM), contains the evaluations of the variables, propositions, etc., i.e., it has the form

$$V([K, L, \{a_{k_i,l_j}\}]) = [K, L, \{V(a_{k_i,l_j})\}] = [K, L, \{\langle \mu_{k_i,l_j}, \nu_{k_i,l_j} \rangle\}]$$

$$= \begin{array}{c|ccccc}
 & l_1 & \cdots & l_j & \cdots & l_n \\
\hline
k_1 & \langle \mu_{k_1,l_1}, \nu_{k_1,l_1} \rangle & \cdots & \langle \mu_{k_1,l_j}, \nu_{k_1,l_j} \rangle & \cdots & \langle \mu_{k_1,l_n}, \nu_{k_1,l_n} \rangle \\
\vdots & \vdots & \ddots & \vdots & \ddots & \vdots \\
k_i & \langle \mu_{k_i,l_1}, \nu_{k_i,l_1} \rangle & \cdots & \langle \mu_{k_i,l_j}, \nu_{k_i,l_j} \rangle & \cdots & \langle \mu_{k_i,l_n}, \nu_{k_i,l_n} \rangle \\
\vdots & \vdots & \ddots & \vdots & \ddots & \vdots \\
k_m & \langle \mu_{k_m,l_1}, \nu_{k_m,l_1} \rangle & \cdots & \langle \mu_{k_m,l_j}, \nu_{k_m,l_j} \rangle & \cdots & \langle \mu_{k_m,l_n}, \nu_{k_m,l_n} \rangle
\end{array},$$

where for every $1 \le i \le m, 1 \le j \le n$: $V(a_{k_i,l_j}) = \langle \mu_{k_i,l_j}, \nu_{k_i,l_j} \rangle$ and $0 \le \mu_{k_i,l_j}$, ν_{k_i,l_j}, $\mu_{k_i,l_j} + \nu_{k_i,l_j} \le 1$.

4 Short Remarks on Intercriteria Analysis

Following [1, 2], here we describe shortly the intercriteria analysis
Let us have an IM

$$A = \begin{array}{c|ccccccc}
 & O_1 & \cdots & O_i & \cdots & O_j & \cdots & O_n \\
\hline
C_1 & a_{C_1,O_1} & \cdots & a_{C_1,O_i} & \cdots & a_{C_1,O_j} & \cdots & a_{C_1,O_n} \\
\vdots & \vdots & \ddots & \vdots & \ddots & \vdots & \ddots & \vdots \\
C_k & a_{C_k,O_1} & \cdots & a_{C_k,O_i} & \cdots & a_{C_k,O_j} & \cdots & a_{C_k,O_n} \\
\vdots & \vdots & \ddots & \vdots & \ddots & \vdots & \ddots & \vdots \\
C_l & a_{C_l,O_1} & \cdots & a_{C_l,O_i} & \cdots & a_{C_l,O_j} & \cdots & a_{C_l,O_n} \\
\vdots & \vdots & \ddots & \vdots & \ddots & \vdots & \ddots & \vdots \\
C_m & a_{C_m,O_1} & \cdots & a_{C_m,O_i} & \cdots & a_{C_m,O_j} & \cdots & a_{C_m,O_n}
\end{array},$$

where for every p, q, $(1 \le p \le m, 1 \le q \le n)$:

- C_p is a criterion, taking part in the evaluation,
- O_q is an object, being evaluated.
- a_{C_p,O_q} is a real number or another object, that is comparable about relation R with the other a-objects, so that for each i, j, k: $R(a_{C_k,O_i}, a_{C_k,O_j})$ is defined. Let $\overline{R}$ be the dual relation of R in the sense that if R is satisfied, then $\overline{R}$ is not satisfied and vice versa. For example, if "R" is the relation "$<$", then $\overline{R}$ is the relation "$>$", and vice versa.

If all numbers $a_{C_p,O_q} \in [0, 1]$, then we can use the intercriteria analysis to detect relations between the criteria, as well as of the relations between objects. But, when there are numbers $a_{C_p,O_q} \notin [0, 1]$, furstly, it must be used the algorithm described in [5].

Let $S_{k,l}^{\mu}$ be the number of cases in which $R(a_{C_k,O_i}, a_{C_k,O_j})$ and $R(a_{C_l,O_i}, a_{C_l,O_j})$ are simultaneously satisfied. Let $S_{k,l}^{\nu}$ be the number of cases in which $R(a_{C_k,O_i}, a_{C_k,O_j})$ and the dual relation $\overline{R}(a_{C_l,O_i}, a_{C_l,O_j})$ are simultaneously satisfied.

Obviously,

$$S_{k,l}^{\mu} + S_{k,l}^{\nu} \leq \frac{n(n-1)}{2}.$$

Now, we introduce new formulas for determining the values of $S_{k,l}^{\mu}$ and $S_{k,l}^{\nu}$. Let for the real number x:

$$sg(x) = \begin{cases} 1, & \text{if } x > 0 \\ 0, & \text{if } x \leq 0 \end{cases}.$$

Therefore, the numbers $S_{k,l}^{\mu}$ and $S_{k,l}^{\nu}$ can be calculated by the following formulas:

$$S_{k,l}^{\mu} = \sum_{i=1}^{n-1} \sum_{j=i+1}^{n} sg(a_{C_k,O_i} - a_{C_k,O_j}).sg(a_{C_l,O_i} - a_{C_l,O_j})$$

$$+ sg(a_{C_k,O_j} - a_{C_k,O_i}).sg(a_{C_l,O_j} - a_{C_l,O_i}),$$

$$S_{k,l}^{\nu} = \sum_{i=1}^{n-1} \sum_{j=i+1}^{n} sg(a_{C_k,O_i} - a_{C_k,O_j}).sg(a_{C_l,O_j} - a_{C_l,O_i})$$

$$+ sg(a_{C_k,O_j} - a_{C_k,O_i}).sg(a_{C_l,O_i} - a_{C_l,O_j}).$$

we see that again

$$S_{k,l}^{\mu} + S_{k,l}^{\nu} \leq \frac{n(n-1)}{2}.$$

Moreover, the present values of $S_{k,l}^{\mu}$ and $S_{k,l}^{\nu}$ coincide with the above ones. Now, for every k, l, such that $1 \leq k < l \leq m$ and for $n \geq 2$, we define

$$\mu_{C_k,C_l} = 2\frac{S_{k,l}^{\mu}}{n(n-1)}, \quad \nu_{C_k,C_l} = 2\frac{S_{k,l}^{\nu}}{n(n-1)}.$$

Therefore, $\langle \mu_{C_k,C_l}, \nu_{C_k,C_l} \rangle$ is an Intuitionistic Fuzzy Pair (IFP, see [3]), i.e., its components and their sum belong to interval [0, 1].

Now, we can construct the IM

$$
\begin{array}{c|ccc}
 & C_1 & \cdots & C_m \\
\hline
C_1 & \langle \mu_{C_1,C_1}, \nu_{C_1,C_1} \rangle & \cdots & \langle \mu_{C_1,C_m}, \nu_{C_1,C_m} \rangle \\
\vdots & \vdots & \ddots & \vdots \\
C_m & \langle \mu_{C_m,C_1}, \nu_{C_m,C_1} \rangle & \cdots & \langle \mu_{C_m,C_m}, \nu_{C_m,C_m} \rangle
\end{array},
$$

that determines the degrees of correspondence between criteria $C_1, \ldots, C_m$.

5 An Example: Relations Among Arithmetic Functions

For the natural number

$$
n = \prod i = 1 p_i^{\alpha_i},
$$

where $k, \alpha_1, \alpha_2, \ldots, \alpha_k \geq 1$ are natural numbers and $p_1, p_2, \ldots, p_k$ are different prime numbers, we define the following functions (for functions φ, ψ, σ see, e.g., [7, 8], for the rest—respectively, [9–11]):

$$
\varphi(n) = \prod_{i=1}^{k} p_i^{\alpha_i - 1}(p_i - 1),
$$

$$
\rho(n) = \prod_{i=1}^{k} (p_i^{\alpha_i} - p_i^{\alpha_i - 1} + \cdots + (-1)^{\alpha_i}),
$$

$$
\psi(n) = \prod_{i=1}^{k} p_i^{\alpha_i - 1}(p_i + 1),
$$

$$
\sigma(n) = \prod_{i=1}^{k} \frac{p_i^{\alpha_i + 1} - 1}{p_i - 1},
$$

$$
\delta(n) = \sum i = 1_k \alpha_i p_1^{\alpha_1} \ldots p_{i-1}^{\alpha_{i-1}} p_i^{\alpha_i - 1} p_{i+1}^{\alpha_{i+1}} \ldots p_k^{\alpha_k},
$$

$$
\zeta(n) = \sum_{i=1}^{k} \alpha_i . p_i.
$$

By definition, for all these functions the following equalities are valid:

$$
\varphi(1) = \rho(1) = \psi(1) = \sigma(1) = \delta(1) = \zeta(1) = 1.
$$

Table 1 Values of the six arithmetic functions

n	$\varphi(n)$	$\rho(n)$	$\psi(n)$	$\sigma(n)$	$\delta(n)$	$\zeta(n)$
2	1	1	3	3	1	2
3	2	2	4	4	1	3
4	2	3	6	7	4	4
5	4	4	6	6	1	5
6	2	2	12	12	5	5
7	6	6	8	8	1	7
8	4	5	12	15	12	6
9	6	7	12	13	6	6
10	4	4	18	18	7	7
11	10	10	12	12	1	11
12	4	6	24	28	16	7
13	12	12	14	14	1	13
14	6	6	24	24	8	8
15	8	8	24	24	8	8
16	8	11	24	31	32	16
17	16	16	18	18	1	17
18	6	7	36	39	21	8
19	18	18	20	20	1	19
20	8	12	36	42	24	9
21	12	12	32	32	10	10
22	10	10	36	36	13	13
23	22	22	24	24	1	23
24	8	10	48	60	44	9
25	20	21	30	31	10	10
26	12	12	42	42	15	15
27	18	20	36	40	27	9
28	12	18	48	56	32	11
29	28	28	30	30	1	29
30	8	8	72	72	31	11
31	30	30	32	32	1	31
32	16	21	48	63	80	10
33	20	20	48	48	14	14
34	16	16	54	54	19	19
35	24	24	48	48	12	12
36	12	21	72	91	60	10
37	36	36	38	38	1	37
38	18	18	60	60	21	21

(continued)

Table 1 (continued)

n	$\varphi(n)$	$\rho(n)$	$\psi(n)$	$\sigma(n)$	$\delta(n)$	$\zeta(n)$
39	24	24	56	56	16	16
40	16	20	72	90	68	11
41	41	41	42	42	1	41
42	12	12	96	96	41	12
43	42	42	44	44	1	43
44	20	30	72	84	48	15
45	24	28	72	78	39	11
46	22	22	72	72	25	25
47	46	46	48	48	1	47
48	16	22	96	124	112	11
49	42	43	56	57	14	14
50	20	21	90	93	45	12

The first values of these functions are given in Table 1.

Now, interpreting objects O_1, O_2, ... as the sequential natural numbers 2, 3, ..., and the criteria C_1, C_2, ... as the values of the six arithmetic functions with parameter these sequential natural numbers, and applying the above described intercriteria analysis, we obtain values $\langle \mu_{C_i,C_j}, \nu_{C_i,C_j} \rangle$ ($1 \leq i, j \leq 7$).

From both tables, we can obtain the list of the pairs of functions that have most near behaviour—see Table 4.

The μ- and ν-values are given in Tables 2 and 3, respectively.

From the point of view of number theory, the values in Table 4 are correct, but here, we use them as a illustration for the successful use of the intercriteria analysis.

The geometrical interpretations of the μ- and ν-values from Table 4 are shown in Fig. 2.

Table 2 μ-values

	n	$\varphi(n)$	$\rho(n)$	$\psi(n)$	$\sigma(n)$	$\delta(n)$	$\zeta(n)$
n	1	0.795068	0.825680	0.857142	0.855442	0.625	0.752551
$\varphi(n)$	0.795068	1	0.914115	0.654761	0.653061	0.439625	0.808673
$\rho(n)$	0.825680	0.914115	1	0.693877	0.703231	0.496598	0.793367
$\psi(n)$	0.857142	0.654761	0.693877	1	0.943877	0.696428	0.643707
$\sigma(n)$	0.855442	0.653061	0.703231	0.943877	1	0.745748	0.633503
$\delta(n)$	0.625	0.439625	0.496598	0.696428	0.745748	1	0.436224
$\zeta(n)$	0.752551	0.808673	0.793367	0.643707	0.633503	0.436224	1

Table 3 ν-values

	n	$\varphi(n)$	$\rho(n)$	$\psi(n)$	$\sigma(n)$	$\delta(n)$	$\zeta(n)$
n	0	0.149659	0.141156	0.092687	0.130952	0.279761	0.213435
$\varphi(n)$	0.149659	0	0.026360	0.241496	0.278061	0.409863	0.110544
$\rho(n)$	0.141156	0.026360	0	0.224489	0.251700	0.375	0.146258
$\psi(n)$	0.092687	0.241496	0.224489	0	0.011054	0.159863	0.282312
$\sigma(n)$	0.130952	0.278061	0.251700	0.011054	0	0.147108	0.320578
$\delta(n)$	0.279761	0.409863	0.375	0.159863	0.147108	0	0.439625
$\zeta(n)$	0.213435	0.110544	0.146258	0.282312	0.320578	0.439625	0

Table 4 List of the pairs of functions that have most near behaviour

f_1	f_2	μ	ν
ψ	σ	0.943877	0.011054
φ	ρ	0.914115	0.026360
n	ψ	0.857142	0.092687
n	σ	0.855442	0.130952
n	ρ	0.825680	0.141156
φ	ζ	0.808673	0.110544
n	φ	0.795068	0.149659
ρ	ζ	0.793367	0.146258
n	ζ	0.752551	0.213435
σ	δ	0.745748	0.147108
ρ	σ	0.703231	0.25170
ψ	δ	0.696428	0.159863
ρ	ψ	0.693877	0.224489
φ	ψ	0.654761	0.241496
φ	σ	0.653061	0.278061
ψ	ζ	0.643707	0.282312
σ	ζ	0.633503	0.320578
n	δ	0.625	0.279761
ρ	δ	0.496598	0.375
φ	δ	0.439625	0.409863
δ	ζ	0.436224	0.439625

Moreover, we can modify the form of the above functions, changing the values of each of them with a natural number n, e.g., $f(n)$ with $\frac{f(n)}{n}$. Then, applying the intercriteria analysis over values from Table 1, we obtain the pair of tables—Tables 5 and 6—that correspond to μ- and ν-values, respectively.

From Tables 5 and 6, we can obtain again the list of the pairs of functions that have most near behaviour—see Table 7.

Fig. 2 Geometrical
interpretation of the results
from Table 4

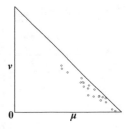

Table 5 μ-values

	$\frac{n}{n}$	$\frac{\varphi(n)}{n}$	$\frac{\rho(n)}{n}$	$\frac{\psi(n)}{n}$	$\frac{\sigma(n)}{n}$	$\frac{\delta(n)}{n}$	$\frac{\zeta(n)}{n}$
$\frac{n}{n}$	1	0.033163	0.002551	0.033163	0.000850	0.000850	0.116496
$\frac{\varphi(n)}{n}$	0.033163	1	0.883503	0.055272	0.055272	0.120748	0.625
$\frac{\rho(n)}{n}$	0.002551	0.883503	1	0.079081	0.159863	0.217687	0.604591
$\frac{\psi(n)}{n}$	0.033163	0.055272	0.079081	1	0.909013	0.836734	0.217687
$\frac{\sigma(n)}{n}$	0.000850	0.055272	0.159863	0.909013	1	0.926020	0.230442
$\frac{\delta(n)}{n}$	0.000850	0.120748	0.217687	0.836734	0.926020	1	0.249149
$\frac{\zeta(n)}{n}$	0.116496	0.625	0.604591	0.217687	0.230442	0.249149	1

Table 6 ν-values

	$\frac{n}{n}$	$\frac{\varphi(n)}{n}$	$\frac{\rho(n)}{n}$	$\frac{\psi(n)}{n}$	$\frac{\sigma(n)}{n}$	$\frac{\delta(n)}{n}$	$\frac{\zeta(n)}{n}$
$\frac{n}{n}$	0	0	0	0	0	0	0
$\frac{\varphi(n)}{n}$	0	0	0.080782	0.944727	0.910714	0.845238	0.230442
$\frac{\rho(n)}{n}$	0	0.080782	0	0.885204	0.836734	0.778911	0.276360
$\frac{\psi(n)}{n}$	0	0.944727	0.885204	0	0.056972	0.129251	0.637755
$\frac{\sigma(n)}{n}$	0	0.910714	0.836734	0.056972	0	0.072278	0.652210
$\frac{\delta(n)}{n}$	0	0.845238	0.778911	0.129251	0.072278	0	0.633503
$\frac{\zeta(n)}{n}$	0	0.230442	0.276360	0.637755	0.652210	0.633503	0

Really, the behaviour of functions φ, σ and ψ is well-known (it is well illustrated in [12]), but their relationship with the three other functions and the relationship betweeh these functions, by the moment is not studied. So, the present research gives a theme of a future research in the area of arithmetic functions.

The geometrical interpretations of the μ- and ν-values from Table 7 are shown in Fig. 3.

Table 7 List of the pairs of functions that have the closed behaviour

f_1	f_2	μ	ν
$\frac{\sigma(n)}{n}$	$\frac{\delta(n)}{n}$	0.92602	0.072278
$\frac{\psi(n)}{n}$	$\frac{\sigma(n)}{n}$	0.909014	0.056972
$\frac{\varphi(n)}{n}$	$\frac{\rho(n)}{n}$	0.883503	0.080782
$\frac{\psi(n)}{n}$	$\frac{\delta(n)}{n}$	0.836735	0.129252
$\frac{\varphi(n)}{n}$	$\frac{\zeta(n)}{n}$	0.625	0.230442
$\frac{\rho(n)}{n}$	$\frac{\zeta(n)}{n}$	0.604592	0.276361
$\frac{\delta(n)}{n}$	$\frac{\zeta(n)}{n}$	0.24915	0.633503
$\frac{\sigma(n)}{n}$	$\frac{\zeta(n)}{n}$	0.230442	0.652211
$\frac{\psi(n)}{n}$	$\frac{\zeta(n)}{n}$	0.217687	0.637755
$\frac{\rho(n)}{n}$	$\frac{\delta(n)}{n}$	0.217687	0.778912
$\frac{\rho(n)}{n}$	$\frac{\sigma(n)}{n}$	0.159864	0.836735
$\frac{\varphi(n)}{n}$	$\frac{\delta(n)}{n}$	0.120748	0.845238
n	$\frac{\zeta(n)}{n}$	0.116497	0
$\frac{\rho(n)}{n}$	$\frac{\psi(n)}{n}$	0.079081	0.885204
$\frac{\varphi(n)}{n}$	$\frac{\sigma(n)}{n}$	0.055272	0.910714
$\frac{\varphi(n)}{n}$	$\frac{\psi(n)}{n}$	0.055272	0.944728
n	$\frac{\psi(n)}{n}$	0.033163	0
n	$\frac{\varphi(n)}{n}$	0.033163	0
n	$\frac{\rho(n)}{n}$	0.002551	0
n	$\frac{\sigma(n)}{n}$	0.000850	0
n	$\frac{\delta(n)}{n}$	0.000850	0

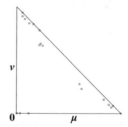

Fig. 3 Geometrical interpretation of the results from Table 7

6 Conclusion

The present research was from one side an illustration of a new intercriteria analysis application, but from another—a possible new direction of research in the area of arithmetic functions, one of the basic areas of number theory.

Acknowledgements The Bulgarian authors are thankful for the support provided by the Bulgarian National Science Fund under Grant Ref. No. DFNI-I-02-5.

References

1. Atanassov, K., Index Matrices: Towards an Augmented Matrix Calculus, Springer, Cham, 2014.
2. Atanassov K., D. Mavrov, V. Atanassova. Intercriteria Decision Making: A New Approach for Multicriteria Decision Making, Based on Index Matrices and Intuitionistic Fuzzy Sets. Issues in Intuitionistic Fuzzy Sets and Generalized Nets, Vol. 11, 2014, 1–8.
3. Atanassov, K., E. Szmidt, J. Kacprzyk, On intuitionistic fuzzy pairs, Notes on Intuitionistic Fuzzy Sets, Vol. 19, 2013, No. 3, 1–13.
4. Atanassov, K., On Intuitionistic Fuzzy Sets Theory, Springer, Berlin, 2012.
5. Atanassov, K., V. Atanassova, P. Chountas, M. Mitkova, E. Sotirova, S. Sotirov, D. Stratiev, Intercriteria analysis over normalized data. In: Proceedings of the 8th IEEE Conference Intelligent Systems, Sofia, 46 September 2016, 136–138.
6. Atanassov, K., V. Atanassova, G. Gluhchev, InterCriteria Analysis: Ideas and problems. Notes on Intuitionistic Fuzzy Sets, Vol. 21, 2015, No. 1, 81–88.
7. Sándor, J., D.S. Mitrinovic, B. Crstici. Handbook of number theory, Vol. I, Springer, Dordrecht, 2006.
8. Nagell T. Introduction to number theory, John Wiley & Sons, New York, 1950.
9. Atanassov K., New integer functions, related to and functions, Bull. of Number Theory and Related Topics Vol. XI (1987), No. 1, 3–26.
10. Atanassov K. On an arithmetic function. Advanced Studies on Contemporary Mathematics, Vol. 8, 2004, No. 2, 177–182.
11. Atanassov, K. A remark on an arithmetic function. Part 2. Notes on Number Theory and Discrete Mathematics, Vol. 15, 2009, No. 3, 21–22.
12. Atanassov, K. Note on φ, ψ and σ-functions. Part 2. Notes on Number Theory and Discrete Mathematics, Vol. 16, 2010, No. 3, 25–28.

Fuzzy Harmony Search Algorithm Using an Interval Type-2 Fuzzy Logic Applied to Benchmark Mathematical Functions

Cinthia Peraza, Fevrier Valdez and Oscar Castillo

Abstract This paper presents a fuzzy harmony search algorithm (FHS) based on an interval type-2 fuzzy logic system for dynamic parameter adaptation. The harmony memory accepting (HMR) and pitch adjustment (PArate) parameters arc changing during the iterations in the improvisation process of this algorithm using the fuzzy system. The FHS has been successfully applied to various benchmark optimization problems. Numerical results reveal that the proposed algorithm can find better solutions when compared to a type-1 FHS and other heuristic methods and is a powerful search algorithm for various benchmark optimization problems.

1 Introduction

The method proposed in this article uses the fuzzy system to change the parameters of the algorithm as the number of iterations advances, this is done in order to achieve an adaptive method and not making the change of these parameters manually, to achieve with this to eliminate the use of fixed parameters. Harmony search is a relatively new metaheuristic optimization algorithm inspired in music and was originally developed by Geem et al. in 2001 [1].

Interval type-2 fuzzy sets (IT2 FSs) and systems [2, 3] have been gaining popularity rapidly in the last decade. The Mendel-John Representation Theorem [4] for IT2 FSs has played an important role. It states that the footprint of uncertainty (FOU) of an IT2 FS is the union of all its embedded type-1 (T1) FSs. This Representation

C. Peraza (✉) · F. Valdez · O. Castillo
Division of Graduate Studies and Research, Tijuana Institute of Technology, 22379 Tijuana, Mexico
e-mail: cinthia_sita@hotmail.com

F. Valdez
e-mail: fevrier@tectijuana.mx

O. Castillo
e-mail: ocastillo@tectijuana.mx

© Springer International Publishing AG, part of Springer Nature 2019   13
M Hadjiski and K T Atanassov (eds.), *Intuitionistic Fuzziness and Other Intelligent Theories and Their Applications*, Studies in Computational Intelligence 757,
https://doi.org/10.1007/978-3-319-78931-6_2

Theorem implies that all these embedded T1 FSs should be considered in deriving new theoretical results for IT2 FSs [5].

In order to improve the fine-tuning characteristics of the HS algorithm, FHS employs a new method based on type-2 fuzzy logic, which is responsible for performing the dynamic adjustment of the harmony memory accepting parameter in the Harmony Search algorithm (HS). The FHS algorithm has the power of the HS algorithm with the fine-tuning feature of mathematical techniques and can outperform either one individually. To show the great power of this method, the FHS algorithm is applied to various standard benchmarking optimization problems. Numerical results reveal that the proposed algorithm is a powerful search algorithm for various optimization problems. In addition, we carefully performed a set of experiments to reveal the impact of the control parameters applied to benchmark mathematical functions and a comparison with other methods: global best harmony search in [6], fuzzy control of parameters to dynamically adapt the HS algorithm for optimization in [7], self-adaptive harmony search algorithm for optimization in [8].

Similarly, there are papers on Harmony Search algorithm applications that use this algorithm to solve real problems. To mention a few, we have the following examples: a parameter setting free harmony search algorithm in [9], a tabu harmony search based approach to fuzzy linear regression in [10], a novel global harmony search algorithm in [11], an improved harmony search algorithm for solving optimization problems [12], a new meta-heuristic algorithm for continuous engineering optimization harmony search theory and practice [13], harmony search algorithms for structural design optimization in [14], and benchmarking of heuristic optimization methods [15]. There are numerous algorithms that have been proposed for solving complex problems. Some of them are special, some are more general. Many problems cannot be solved by deterministic algorithms, and heuristic algorithms can also be used. In the set of Meta heuristic algorithms, we can find: A fuzzy PSO and GA [16], which proposed a type-1 fuzzy system for parameter adaptation of PSO and GA, using $c1$ and $c2$ as variables to adjust through the iterations.

Some existing works on Fuzzy parameter adaptation in optimization are: Some neural net training examples [17], which proposed a type 1 fuzzy system for parameter adaptation of back propagation. A differential evolution algorithm in [18], which proposed a type-1 fuzzy system for parameter adaptation of DE, using the F and C parameters as variables to adjust through the iterations.

The remainder of the paper is organized as follows: Sect. 2 shows the equations and the basic concepts of the harmony search algorithm. Section 3 describes the proposed method. Section 4 shows the methodology for parameter adaptation. Section 5 shows comparison of results of type-2 FHS and type-1 FHS applied to benchmarking optimization problems. Finally, Sect. 6 describes the conclusions.

2 Harmony Search Algorithm

The Harmony search (HS) algorithm was recently developed in an analogy with music improvisation process where music players improvise the pitches of their instruments to obtain better harmony [13]. The steps in the procedure of harmony search are shown in Fig. 1. They are as follows:

Step 1 Initialize the problem and algorithm parameters.
Step 2 Initialize the harmony memory.
Step 3 Improvise a new harmony.
Step 4 Update the harmony memory.
Step 5 Check the stopping criterion.

A. *These steps are described in the next five subsections.*
Initialize the problem and algorithm parameters
In Step 1, the optimization problem is specified as follows:

Minimize
$$f(x) \text{ subject to } x_i \in X_i = 1, 2, \ldots, N \tag{1}$$

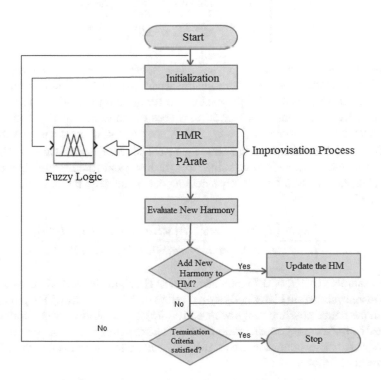

Fig. 1 Variation of HMR using the type 2 fuzzy system of iteration number

where $f(x)$ is an objective function; x is the set of each decision variable x_i; N is the number of decision variables (dimensions), X_i is the set of the possible range of values for each decision variable, that is $L_{x_i} \leq X_i \leq U_{x_i}$ and L_{x_i} and U_{x_i} are the lower and upper bounds for each decision variable. The HS algorithm parameters are also specified in this step. These are the harmony memory size (*HMS*), or the number of solution vectors in the harmony memory; harmony memory considering rate (*HMR*), pitch adjusting rate (*PArate*); and the number of iterations (NI), or stopping criterion. The harmony memory (*HM*) is a memory location where all the solution vectors (set of decision variables) are stored. This HM is similar to the genetic pool in the GA. Here, *HMR* and *PArate* are used to improve the solution vector. Booth are defined in step 3.

In step 2, the *HM* matrix is filled with as many randomly generated solution vectors as the *HMS*

$$HM = \begin{bmatrix} x_1^1 & x_2^1 & \cdots & x_{N-1}^1 & x_N^1 \\ x_1^2 & x_2^2 & \cdots & x_{N-1}^2 & x_N^2 \\ \vdots & \vdots & \vdots & \vdots & \vdots \\ x_1^{HMS-1} & x_2^{HMS-1} & \cdots & x_{N-1}^{HMS-1} & x_N^{HMS-1} \\ x_1^{HMS} & x_2^{HMS} & \cdots & x_{N-1}^{HMS} & x_N^{HMS} \end{bmatrix} \qquad (2)$$

A new harmony vector, $x' = (x_1', x_2', \ldots, x_N')$, is generated based on three rules: (1) memory consideration, (2) pitch adjustment and (3) random selection. Generating a new harmony that is called 'Improvisation'. In the harmony consideration, the value of the first decision variable (x_1') for the new vector is chosen from any of the values in the specified *HM* range $(x_1^{'1} - x_1^{'HMS})$. Values of the other decision variables $(x_1', x_2', \ldots, x_N')$ are chosen in the same manner. The *HMR*, which varies between 0 and 1, is the rate of choosing one value from the historical values stored in the *HM*, while (1-*HMR*) is the rate randomly selecting one value from the possible range of values [19].

$$x_1' \leftarrow \begin{cases} x_1' \in \{x_1', x_2', \ldots, x_i^{'HMS}\} \; with \; probability \; HMR, \\ \overline{x_1' \in X_i \; with \; probability \; (1 - HMR)} \end{cases} \qquad (3)$$

For example, a *HMR* of 0.75 indicates that the HS algorithm will choose the decision variable value from historically stored values in the *HM* with an 75% probability or from the entire possible range with a (100–75)% probability. Every component obtained by the memory consideration is examined to determine whether it should be pitch-adjusted. This operation uses the *PArate* parameter, which is the rate of pitch adjustment as follows:

$$Pitch \; adjusting \; decision \; for \; x_1' \leftarrow \left\{ \frac{Yes \; probability \; PArate}{No \; with \; probability (1 - PArate)} \right\} \qquad (4)$$

The value of $(1 - PArate)$ sets the rate of doing nothing. If the pitch adjustment decision for x_1' is YES, x_1' is replaced as follows:

$$x_1' \leftarrow \pm rand() * bw, \tag{5}$$

where

bw is an arbitrary distance bandwidth
Rand () is a random number between 0 and 1

In step 3, *HMR* (harmony memory accepting), PArate (pitch adjustment) or random selection is applied to each variable of the New Harmony vector in turn.

In step 4, the new harmony vector, $x' = (x_1', x_2', \ldots, x_N')$, is better than the worst harmony in the *HM*, judged in terms of the objective function value, the new harmony is included in the *HM* and the existing worst harmony is excluded from the *HM*.

In step 5, if the stopping criterion (maximum number of iterations) is satisfied, computation is terminated. Otherwise, steps 3 and 4 are repeated.

3   Proposed Method

This section explains the proposed type-2 fuzzy harmony search (FHS) algorithm. A brief overview of the modification procedures of the proposed FHS algorithm is presented. The HMR and PArate parameters are introduced in the type-2 fuzzy system to help the algorithm to globally and locally find improved solutions, respectively. The HMR and PArate in the HS algorithm are very important parameters in fine-tuning the optimized solution vectors, and can be potentially useful in improving the convergence rate of the algorithm in finding optimal solutions. So fine adjustment of these parameters is of great interest. The original HS algorithm uses fixed values for both HMR and PArate. In the HS method the HMR and PArate values are adjusted in the initialization step and cannot be changed during new iterations. The main drawback of this method appears in the number of iterations the algorithm needs to find an optimal solution. If small HMR values are selected, only the best harmonies and may converge very slowly. Furthermore, for large HMR values almost all the harmonies are used in memory of harmony, then other harmonies are not well explored, leading to potentially erroneous solutions.

Small PArate values can cause poor performance of the algorithm and considerable increase in the iterations needed to find optimum solution. Although small values in final iterations increase the fine-tuning of solution vectors, but in early iterations must take a bigger value to enforce the algorithm to increase the diversity of the solution vectors. The HMR parameter represents the exploitation in the space search of the algorithm. The PArate parameter represents the exploration in space search of the algorithm.

The main difference between the type-2 FHS and traditional HS method is in the form of adjusting HMR and PArate, the type-2 FHS algorithm uses the variable HMR and PArate in improvisation step. FHS uses a type-2 fuzzy system to be responsible of dynamically changing the HMR and PArate parameters in the range from 0 to 1 in each iteration number.

Currently algorithms use techniques to move its parameters such as linear equations, methods based on probability, hybridization with other existing algorithms, all this in order to improve the performance of the algorithm, in our case we use a fuzzy system to make it responsible for moving these parameters in a dynamic fashion based on the iterations. We decided to apply type 2 fuzzy logic in this case because in previous works we use type 1 and it gave us good results, by using type-2 fuzzy systems can handle more uncertainty in the range that we determine and seek the stability of the method for this we decided to try other mathematical functions using different number of dimensions for give it more complexity to the problem and to obtain significant evidence as shown in Fig. 1 and expressed as follows.

4   Methodology for Parameter Adaptation

In Sects. 2 and 3 we show the most important parameters of the algorithm, based on the literature, so we decided to use the HMR and PArate parameter as the outputsfor the fuzzy system and must be in the range of 0–1, plus it is also suggested that changing the parameter HMR and PArate dynamically during the execution of this algorithm can produce better results.

In addition, it is also found that the algorithm Performance measures, such as: the iterations, need to be considered to run the algorithm, among others. In our work all the above are taken in consideration for designing the fuzzy systems that modify the harmony memory accepting and PArate parameters for dynamically changing these parameters in each iteration of the algorithm. For measuring the iterations of the algorithm, it was decided to use a percentage of iterations, i.e. when starting the algorithm, the iterations will be considered "low", and when the iterations are completed they will be considered "high" or close to 100%. The design of the fuzzy system can be found in Figs. 2 and 3 and each contains one input and one output. To represent this idea, we use [20]:

$$Iteration = \frac{Current\ Iteration}{Maximun\ of\ iterations} \tag{6}$$

The design of the input variable for each fuzzy system can be appreciated in Fig. 4, which shows the iteration input. This input is granulated into three triangular membership functions. For the output variables, as mentioned above, the

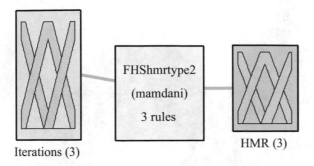

Fig. 2 Fuzzy System for HMR parameter adaptation

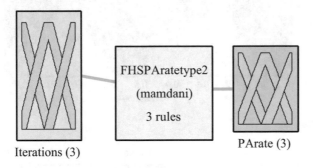

Fig. 3 Fuzzy System for PArate parameter adaptation

recommended values for HMR and PArate are between 0 to 1, so that the output variables were designed using this range of values. The output is granulated into three triangular membership functions. The design of the output variables can be found in Figs. 5 and 6.

The surface of the fuzzy system shows that the HMR parameter changes dynamically with the increasing iterations shown in Fig. 7 and the surface of the fuzzy system shows that the PArate parameter changes dynamically with decreasing iterations shown in Fig. 8.

Based on the behavior of the HMR parameter we decide to use rules in an increasing fashion to first explore the search space and eventually exploit this in order to find better solutions the rules for the fuzzy system are shown in Fig. 9. For the PArate parameter we decide to use rules in a decreasing fashion to first exploit the search space and eventually explore this in order to find better solutions the rules for the fuzzy system are shown in Fig. 10.

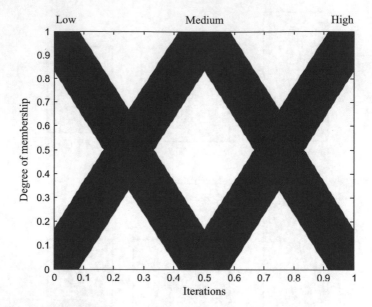

Fig. 4 Input: Iteration

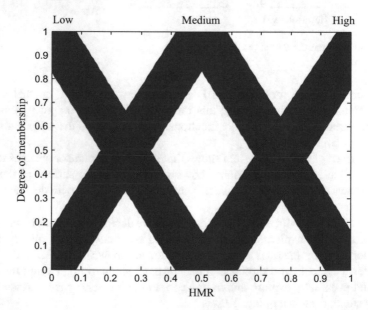

Fig. 5 Output: HMR

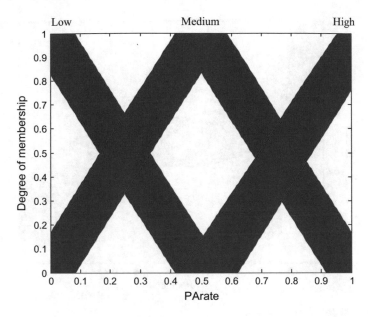

Fig. 6 Output: PArate

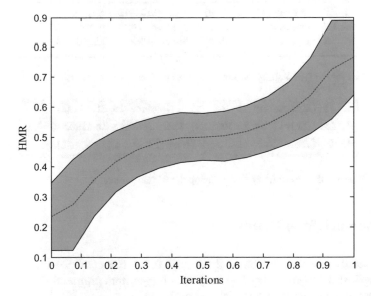

Fig. 7 Surface for the HMR parameter

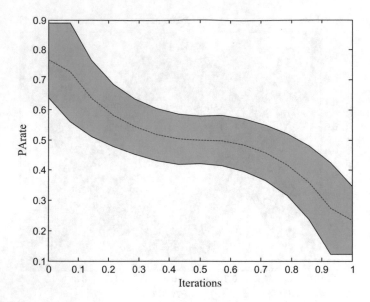

Fig. 8 Surface for PArate parameter

```
1. If (Iteration is Low) then (HMR is Low) (1)
2. If (Iteration is Medium) then (HMR is Medium) (1)
3. If (Iteration is High) then (HMR is High) (1)
```

Fig. 9 Rules for the type-2 fuzzy system using the HMR parameter

```
1. If (Iterations is Low) then (PArate is High) (1)
2. If (Iterations is Medium) then (PArate is Medium) (1)
3. If (Iterations is High) then (PArate is Low) (1)
```

Fig. 10 Rules for the type-2 fuzzy system using the PArate parameter

5   Comparison of Results

In this section the comparison of the type-2 fuzzy Harmony Search algorithm is made against the type-1 fuzzy harmony search algorithm proposed in [7] and other methods for benchmarking optimization problems.

We decided to use different dimensions in each mathematical function to give more complexity to the problem, for example if we use 30 dimensions we are using the same number of dimensions in the space to make the problem more complex. We decided to apply type 2 fuzzy logic to give more complexity to the algorithm and make it stable the more complex is a problem must obtain better results with respect to type-1 fuzzy logic.

Table 1 Comparison against Peraza et al. [7]

Function	Type-1 FHS [7]	Type-2 FHSHMR
Sphere	6.33E-07	1.44E-10
Rosenbrock	1.36E-08	2.52E-12
Ackley	7.87E-05	9.28E-09
Rastrigin	5.28E-05	9.55E-06
Zakharov	1.25E-04	1.41E-05
Sum squared	3.72E-06	2.09E-06

Table 2 Shows the results that were obtained using type-1 fuzzy harmony search and type-2 fuzzy harmony search with dynamic adaptation the PArate parameter

Function	Type-1 FHS [7]	Type-2 FHS PArate
Sphere	4.33E-07	1.27E-10
Rosenbrock	1.20E-08	3.42E-04
Ackley	6.58E-05	6.50E-09
Rastrigin	4.28E-05	1.05E-08
Zakharov	2.15E-04	2.30E-03
Sum squared	8.72E-06	7.43E-06

The first comparison is against Peraza et al. [7], where they proposed a new way for the harmony memory to automatically adjust parameter values using a type-1 fuzzy system with dynamic adaptation the HMR parameter. To make a fair comparison, we use the same constraints used in [7], this is we use the same parameters for harmony memory, iterations, dimensions, functions and range. Table 1 summarizes the results presented in [7], and we also include the results using the proposed approach in this paper.

Table 1 shows the results that were obtained using the type-1 fuzzy harmony search and type-2 fuzzy harmony search with dynamic adaptation the HMR parameter.

In Table 1 the Type-1 FHS and type-2 FHS algorithms are applied to various standard benchmarking optimization problems. The result is the average of the 30 experiments with the parameters used in the method [7], we can notice the overall averages obtained in each mathematical function for each method, and the best results are with our proposed method. The plot of the experiments for each method simple harmony search algorithm (HS) and fuzzy harmony search algorithm (FHS) for all functions are shown in Fig. 11 for type-1 and Fig. 12 for type-2, respectively.

In Table 2 the Type 1 FHS and type-2 FHS algorithms with dynamic adaptation of the PArate parameter are applied to various standard benchmarking optimization problems. The result is the average of the 30 experiments with the parameters used in the method [7], we can notice the overall averages obtained in each mathematical function for each method, and the best results are with our proposed method. The plot of the experiments for each method simple harmony search algorithm (HS) and fuzzy harmony search algorithm (FHS) for all functions is shown in Fig. 12.

The second comparison is against Wang et al. [8], where they proposed a new way for the harmony memory to automatically adjust parameter values.

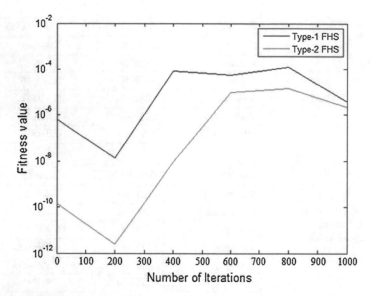

Fig. 11 Plot of the experiments of the type-1 FHS (Blue Line) and type-2 FHS (Green Line), Dim = 10. All results have been averaged over 30 runs

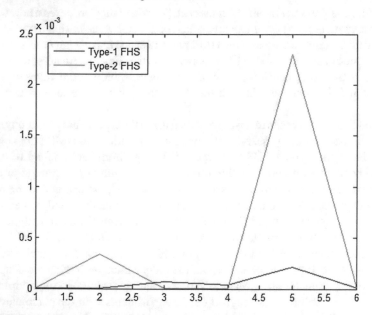

Fig. 12 Plot of the experiments of the type-1 FHS (Blue Line) and type-2 FHS (Green Line), Dim = 10. All results have been averaged over 30 runs

In addition, the pseudo-random number generator is also replaced by the low-discrepancy sequences for initialization of the harmony memory. We use the same

Table 3 Comparison with Wang et al. results in 30 dimensions [8]

30 dimensions					
HS50 [8]			Our approach		
Sphere	MEAN	6.92E-07	Sphere	MEAN	3.52E-10
	S.D	1.10E-06		S.D	3.85E-10
Rosenbrock	MEAN	2.65E+01	Rosenbrock	MEAN	3.12E-10
	S.D	5.68E-01		S.D	3.42E-10
Ackley	MEAN	7.81E-04	Ackley	MEAN	8.47E-07
	S.D	4.66E-04		S.D	1.14E-06
Griewank	MEAN	8.45E-05	Griewank	MEAN	4.58E-03
	S.D	2.38E-04		S.D	2.11E-02

Table 4 Comparison with Wang et al. results in 100 dimensions [8]

100 dimensions					
HS50 [8]			Our approach		
Sphere	MEAN	1.53E-02	Sphere	MEAN	3.79E-10
	S.D	1.18E-02		S.D	4.74E-10
Rosenbrock	MEAN	9.60E+01	Rosenbrock	MEAN	4.07E-10
	S.D	5.40E-01		S.D	4.01E-10
Ackley	MEAN	2.92E-02	Ackley	MEAN	7.46E-07
	S.D	9.81E-03		S.D	8.50E-07
Griewank	MEAN	6.66E-03	Griewank	MEAN	1.40E-02
	S.D	3.81E-03		S.D	3.63E-02

constrains used in [8], and in Tables 3 and 4 the results from [8] are presented, compared with the proposed approach in this paper.

From Table 3 can be appreciated the mean and standard deviations of the benchmark function optimization results in 30 dimensions for each mathematical function, we use the same constrains used in [8]. The results were averaged over 30 runs. We can note that as the problem gets more complex the type-2 algorithm sometimes get better results. Tables 4 shows the simulation results for the 100 dimensions in each function.

From Table 4 can be appreciated that mean and standard deviation of the benchmark function optimization results in 100 dimensions. The results were averaged over 30 runs.

The third comparison is against Mahamed et al. [6], where they proposed a concept from swarm intelligence to enhance the performance of the HS. We use the same constrains used in [6], and in Tables 5 and 6 the results from [6] are presented, compared with the proposed approach in this paper.

From Table 5 can be appreciated the mean and standard deviations of the benchmark function optimization results in 30 dimensions, only in the six hump camel backfunction we used 2 dimensions. The results were averaged over 30 runs.

Table 5 Comparison with Mahamed et al. results in 30 dimensions [6]

30 dimensions					
GHS [6]			Our approach		
Sphere	MEAN	1.00E-05	Sphere	MEAN	3.42E-10
	S.D	2.20E-05		S.D	3.64E-10
Rosenbrock	MEAN	4.97E+01	Rosenbrock	MEAN	4.54E-03
	S.D	5.92E+01		S.D	2.49E-02
Ackley	MEAN	2.09E-02	Ackley	MEAN	5.23E-07
	S.D	2.17E-02		S.D	4.57E-07
Griewank	MEAN	1.02E-01	Griewank	MEAN	4.60E-04
	S.D	1.76E-01		S.D	1.06E-03
Schwefel's	MEAN	7.82e-02	Schwefel's	MEAN	1.21e+00
	S.D	1.14e-01		S.D	1.01e-01
Rastrigin	MEAN	8.63e-03	Rastrigin	MEAN	4.29e-06
	S.D	1.53e-02		S.D	4.42e-06
Hyper-ellipsoid	MEAN	5.15e+03	Hyper-ellipsoid	MEAN	8.86e-06
	S.D	6.35e+03		S.D	8.61e-06
Six hump camel back	MEAN	−1.03e+00	Six hump camel back	MEAN	-1.03e+00
	S.D	1.80e-05		S.D	6.78e-16

Table 6 Comparison with Mahamed et al. results in 100 dimensions [6]

100 dimensions					
GHS [6]			Our approach		
Sphere	MEAN	2.23E+00	Sphere	MEAN	3.23E-10
	S.D	5.65E-01		S.D	4.56E-10
Rosenbrock	MEAN	2.60E+06	Rosenbrock	MEAN	1.80E-03
	S.D	9.16E+05		S.D	9.86E-03
Ackley	MEAN	8.77E+00	Ackley	MEAN	1.02E-06
	S.D	8.80E-01		S.D	1.48E-06
Griewank	MEAN	5.43E+01	Griewank	MEAN	1.42E-02
	S.D	1.86E+01		S.D	3.78E-02
Schwefel's	MEAN	1.90e+01	Schwefel's	MEAN	3.62e+03
	S.D	5.09e+00		S.D	4.76e+02
Rastrigin	MEAN	8.07e+01	Rastrigin	MEAN	4.56e-06
	S.D	3.04e+01		S.D	7.22e-06
Hyper-ellipsoid	MEAN	3.22e+05	Hyper-ellipsoid	MEAN	7.01e-06
	S.D	3.96e+04		S.D	7.75e-06

From Table 6 can be appreciated the mean and standard deviations of the benchmark function optimization results in 100 dimensions. The results were averaged over 30 runs. We can note that as the more complex the problem the type-2 sometimes get better results.

6 Conclusions

In this paper, we proposed a new fuzzy harmony search with dynamic adjustment of parameters using a type-2 fuzzy system.

A comparison with four methods was performed namelythe type-1 fuzzy HS, self-adaptive harmony search algorithm, global best harmony search, applied to 4, 6 and 8 benchmark mathematical functions using 10, 30 and 100 dimensions.

The goal of this work was to extend the type-1–type-2 fuzzy system to validate the operation of the in some functions, it went very well as the type-2 fuzzy systems support a higher level of uncertainty and complexity level.

In this work the dynamic adjustment of two internal parameters of the HMR algorithm that represents the operation of the algorithm and PArate that represents the exploration of the algorithm is presented, the type-2 fuzzy system of this last parameter PArate was applied to mathematical functions of reference, which we can to conclude in this work is that when using the adjustment of the PArate parameter it is possible to obtain better results with respect to the average in the functions sphere, Ackley, Rastrigin, sum squared. Better results are achieved by adapting the HMR parameter than the PArate, based on this comparison are presented with other methods using HMR.

Based on these observations, the proposed method was compared with the self-adaptive harmony search algorithm and other methods. The numerical results indicated that the proposed method offers much higher performance to update the existing methods on three optimization problems. Besides, it is interesting to note that in original methods can get better performance obtained with the parameters from the experiments. Testing with more mathematical functions and applying the method with higher number of dimensions using type-2 fuzzy logic because by using such systems we can find greater advantage of the level of uncertainty and robustness.

As future work we want to apply a type-2 fuzzy system to dynamically adjust the all parameters in this algorithm and apply a second input to achieve better results and make a combination of the parameters in the output of the fuzzy system.

Acknowledgements We would like to express our gratitude to the CONACYT and Tijuana Institute of Technology for the facilities and resources granted for the development of this research.

References

1. Geem Z.: Music inspired harmony Search Algorithm theory and applications, Studies in computational intelligence, pp. 8–121, Springer, Heidelberg, Germany (2009).
2. Mendel J. M., Uncertain Rule-Based Fuzzy Logic Systems: Introduction and New Directions. Upper Saddle River, NJ: Prentice-Hall, 2001.
3. Zadeh L. A., "The concept of a linguistic variable and its application to approximate reasoning-1," Information Sciences, vol. 8, pp. 199–249, 1975.
4. Mendel J. M. and R. I. John, "Type-2 fuzzy sets made simple," IEEE Trans. on Fuzzy Systems, vol. 10, no. 2, pp. 117–127, 2002.
5. Mendel J. M., "On answering the question "where do i start in order to solve a new problem involving interval type-2 fuzzy sets?"," Information Sciences, vol. 179, no. 19, pp. 3418–3431, 2009.
6. Mahamed G., Mahdavi M.: Global best harmony search, Applied Mathematics and Computation, pp. 1–14. Elsevier, Amsterdam, Holland (2008).
7. Peraza C., Valdez F., Castillo O.: Fuzzy control of parameters to dynamically adapt the hs algorithm for optimization, Fuzzy Information Processing Society (NAFIPS) held jointly with 2015 5th World Conference on Soft Computing, pp. 1–6, IEEE (2015).
8. Wang C., Huang Y.: Self adaptive harmony search algorithm for optimization, Expert Systems with Applications Volume 37, pp. 2826–2837, Elsevier (2010).
9. Geem Z., Sim K., Parameter setting free harmony search algorithm, Applied Mathematics and Computation, pp. 3881–3889. Elsevier, Chung Ang, China (2010).
10. Hadi M., Mehmet A., Mashinchi M., Pedrycz W.: A Tabu Harmony Search Based Approach to Fuzzy Linear Regression, Fuzzy Systems, IEEE Transactions on, pp. 432–448. IEEE, New Jersey, USA (2011).
11. Dexuan Z., Yanfeng G., Liqun G., Peifeng W.: A Novel Global Harmony Search Algorithm for Chemical Equation Balancing, In Computer Design and Applications (ICCDA), 2010 International Conference on Vol. 2, pp. V2–1. IEEE. (2010).
12. Mahdavi M., Fesanghary M., Damangir E.: An improved harmony search algorithm for solving optimization problems, applied Mathematics and Computation, pp. 1567–1579. Elsevier, Amsterdam, Holland (2007).
13. Geem Z., Lee K.: A new meta-heuristic algorithm for continuous engineering optimization harmony search theory and practice, Computer methods in applied mechanics and engineering, pp. 3902–3933. Elsevier, Maryland, USA (2004).
14. Geem Z.: Harmony search algorithms for structural design optimization, Studies in computational intelligence, Vol. 239, pp. 8–121. Springer, Heidelberg, Germany (2009).
15. Štefek A.: Benchmarking of heuristic optimization methods, Mechatronika 14th International Symposium, pp. 68–71, IEEE (2011).
16. Valdez F., Melin P., Castillo O.: Fuzzy Control of Parameters to Dynamically Adapt the PSO and GA Algorithms, Fuzzy Systems International Conference, pp. 1–8, IEEE, Barcelona, Spain (2010).
17. Arabshahi, Payman, et al. "Fuzzy parameter adaptation in optimization: Some neural net training examples." Computing in Science & Engineering 1 (1996): 57–65.
18. Ochoa P., Castillo O., Soria J., Differential evolution with dynamic adaptation of parameters for the optimization of fuzzy controllers, Recent Advances on Hybrid Approaches for designing intelligent systems, pp. 275–288. Springer, Heidelberg, Germany (2013).
19. Yang X.: Nature Inspired Metaheuristic Algorithms, Second Edition, University of Cambridge, United Kingdom, pp. 73–76, Luniver Press (2010).
20. Olivas F., Melin P., Castillo O., et al., Optimal design of fuzzy classification systems using PSO with dynamic parameter adaptation through fuzzy logic, pp. 2–11, Elsevier (2013).

Mixture Initialization Based on Prior Data Visual Analysis

Evgenia Suzdaleva and Ivan Nagy

Abstract The initialization is known to be a critical task for running a mixture estimation algorithm. A majority of approaches existing in the literature are related to initialization of the expectation-maximization algorithm widely used in this area. This study focuses on the initialization of the recursive mixture estimation for the case of normal components, where the mentioned methods are not applicable. Its key part is a choice of the initial statistics of normal components. Several initialization techniques based on visual analysis of prior data are discussed. Validation experiments are presented.

1 Introduction

The initialization is known to be a critical task for running a mixture estimation algorithm. Mixture models are often used for description of multi-modal systems, whose behavior can switch among different working modes. Such modeling is demanded in a variety of application areas, including, e.g., fault detection (fault or non-fault mode), car diagnostics (eco-driving or sport mode, etc.), traffic flow control (the level of service), big data issues, etc., see, for instance, [1–3].

The mixture model consists of several components that describe the individual working modes of the observed system and of their switching model. The last is considered as the random Markov process called the pointer [4, 5], and its value at the corresponding time instant indicates the currently active component (i.e., the working mode). In reality, parameters of neither the components nor the pointer model are

E. Suzdaleva (✉) · I. Nagy
Department of Signal Processing, The Czech Academy of Sciences,
Institute of Information Theory and Automation, Pod vodárenskou věží 4,
18208 Prague, Czech Republic
e-mail: suzdalev@utia.cas.cz

I. Nagy
Faculty of Transportation Sciences, Czech Technical University, Na Florenci 25,
11000 Prague, Czech Republic
e-mail: nagy@utia.cas.cz

© Springer International Publishing AG, part of Springer Nature 2019   29
M Hadjiski and K T Atanassov (eds.), *Intuitionistic Fuzziness and Other Intelligent Theories and Their Applications*, Studies in Computational Intelligence 757,
https://doi.org/10.1007/978-3-319-78931-6_3

available. Thus the mixture estimation problem consists, in general, in estimation of the component and the pointer model parameters, and also in the pointer value estimation.

The mixture estimation approaches found in the literature are mainly based on (i) the iterative expectation-maximization (EM) algorithm [6], see, e.g., [7, 8]; (ii) the approximative Variational Bayes approach [9, 10]; (iii) sampling Markov Chain Monte Carlo techniques, e.g., [11–13]. Closely related tasks are also discussed in [14, 15].

A different non-numerical approach is given by the recursive Bayesian estimation theory for static mixtures [4, 5], individual normal components [16] and dynamic mixtures [17], which, unlike the above mentioned sources, represent on-line data-based estimation algorithms avoiding numerical iterative computations. The present research project supports their philosophy in developing the mixture estimation algorithms.

The mixture estimation algorithm should be initialized before starting. In the considered context the initialization primarily lies in specifying (i) distributions of components, (ii) the number of components, and (iii) the prior probability density functions (pdfs) describing parameters of components and of the pointer model. The present paper is limited by mixtures of normal components.

A series of papers was found in the area of the mixture initialization. For instance, the paper [18] proposes the initialization of the EM algorithm via a strategy defining mean vectors by choosing points with higher concentrations of neighbors. It uses a truncated normal distribution for the preliminary estimation of covariance matrices.

Another paper [19] describes a new method for random initialization of the EM algorithm based on selecting the feature vector from a set of candidate vectors, located farthest from already initialized components. The Mahalanobis distance is used. The paper [20] is devoted to simple and fast approaches of the initialization of the EM algorithm based on the well-known clustering algorithms. The paper [21] proposes the EM initialization method by partition of the training set to be modeled individually by single experts and the subsequent initialization of models on a partition subset. The paper [22] initializes a mixture via the EM algorithm using a product kernel estimate of pdfs and the gradient method for local extrema finding.

It is seen that the majority of studies is primarily oriented at application of the EM algorithm. Under the adopted theory [4, 5, 16, 17] not using the EM algorithm, the initialization focuses on the number of components and the initial statistics of the Gauss-inverse-Wishart parameter pdfs. In this field the paper [23] is found, which (applied to the presented subproblem) leads to weighting the initial statistics of the parameter pdfs.

The present paper is the extended version of the work [24]. It considers the initialization primarily based on detecting the initial centers of components via the visualization analysis of the prior or expert knowledge. In the case of static normal components this expert-based procedure is rather effective. For dynamic mixture components the task is more complicated. The paper considers several ways of initialization of dynamic components: (i) fixation of covariance matrices; (ii) imitation of the static case; (iii) repeated use of the data sample, see, e.g., [4]; and (iv) weighting the initial statistics [23], and validates them experimentally on real data. The paper demonstrates that a relatively small amount of prior data used for the mixture initialization contributes to a faster stabilization of parameter estimates during the on-line estimation.

The paper is organized in the following way. Section 2 introduces models. Section 3 gives necessary basic facts about the mixture estimation algorithm and specifies the initialization problem. Section 4 describes the mentioned initialization approaches. Section 5 provides results of experiments. Conclusions and open problems are given in Sect. 6.

2 Models

Let's consider a multi-modal system, which at each discrete time instant $t = 1, 2, \ldots$ generates the continuous data vector y_t. It is assumed that the observed system works in m_c working modes, each of them is indicated at each time instant t by the value of the unmeasured dynamic discrete variable $c_t \in \{1, 2, \ldots, m_c\}$, which is called the pointer [5].

The observed system is supposed to be described by a mixture model, which (in this paper) consists of m_c components. The components can be represented either by

$$\text{the static pdf } f\left(y_t | \Theta, c_t = i\right), \ \forall i \in \{1, 2, \ldots, m_c\}, \tag{1}$$

$$\text{or by the dynamic pdf } f\left(y_t | \psi_{t-1}, \Theta, c_t = i\right), \tag{2}$$

where Θ is a collection of parameters of all components, and $\Theta \equiv \{\Theta_i\}_{i=1}^{m_c}$, where Θ_i includes parameters of the i-th component in the sense that $f\left(y_t | \Theta, c_t = i\right) = f\left(y_t | \Theta_i\right)$ for $c_t = i$, and $\psi_{t-1} = [y_{t-1}, y_{t-2}, \ldots, y_{t-n}]'$ is the regression vector with the memory length n.

This paper focuses on using the pdfs (1) or (2) with the normally distributed white noise. In this case the pdfs are specified as follows.

2.1 Static Components

The pdf (1) has the form $\forall i \in \{1, 2, \ldots, m_c\}$

$$(2\pi)^{-N/2} |r_i|^{-1/2} \exp\left\{-\frac{1}{2}[y_t - \theta_i]'r_i^{-1}[y_t - \theta_i]\right\}, \tag{3}$$

where N denotes a dimension of the vector y_t; θ_i represents the center of the i-th component; r_i is the covariance matrix of the involved normal noise, which defines the shape of the component (i.e., in the case of the diagonal r_i the component is round-shaped), and $\Theta_i \equiv \{\theta_i, r_i\}$.

2.2 Dynamic Components

The pdf (2) is specified as

$$(2\pi)^{-N/2}|r_i|^{-1/2}\exp\left\{-\frac{1}{2}[y_t-\theta_i\psi_{t-1}]'r_i^{-1}[y_t-\theta_i\psi_{t-1}]\right\}, \tag{4}$$

where unlike (1) the parameter θ_i is a collection of regression coefficients of the i-th component, whose number corresponds to the memory length n used for the regression vector ψ_{t-1}. A rest of notations are identical to the previous case.

2.3 Dynamic Pointer Model

Switching the active components, either (1) or (2), is described by the dynamic model

$$f(c_t=i|c_{t-1}=j,\alpha),\ i,j\in\{1,2,\ldots,m_c\}, \tag{5}$$

which is represented by the transition table

	$c_t=1$	$c_t=2$	$\cdots$	$c_t=m_c$
$c_{t-1}=1$	$\alpha_{1|1}$	$\alpha_{2|1}$	$\cdots$	$\alpha_{m_c|1}$
$c_{t-1}=2$	$\alpha_{1|2}$		$\cdots$	
$\cdots$	$\cdots$	$\cdots$	$\cdots$	$\cdots$
$c_{t-1}=m_c$	$\alpha_{1|m_c}$		$\cdots$	$\alpha_{m_c|m_c}$

where the parameter α is the $(m_c\times m_c)$-dimensional matrix, and its entries $\alpha_{i|j}$ are non-negative probabilities of the pointer $c_t=i$ (expressing that the i-th component is active at time t) under condition that the previous pointer $c_{t-1}=j$.

3 Recursive Mixture Estimation

Formulation of the initialization problem requires a preliminary outline of the recursive approach to the Bayesian mixture estimation. The algorithm to be effectively initialized is based on the paper [5], which proposes the solution for normal mixtures with the static pointer model, and on [17] considered the problem for the dynamic pointer model. In the context of the introduced mixture of components (1) or (2) and of the pointer model (5), the estimation problem concerns the unknown parameters Θ and α and the pointer value c_t. Derivations are based on construction of the joint pdf of all variables to be estimated and application of the Bayes rule and of the chain rule, see e.g., [16]. Here they are outlined briefly to present the necessary theoretical background for static components (1) with a subsequent explanation of changes in the case of using (2).

Assuming that Θ and α, and y_t and α, and c_t and Θ are mutually independent, and denoting the data collection $y(t) = \{y_0, y_1, \ldots, y_t\}$, where y_0 stands for prior data, the joint pdf of all variables to be estimated has the form $\forall i, j \in \{1, 2, \ldots, m_c\}$

$$\underbrace{f(\Theta, c_t = i, c_{t-1} = j, \alpha | y(t))}_{joint\ posterior\ pdf} \tag{6}$$

$$\propto \underbrace{f(y_t, \Theta, c_t = i, c_{t-1} = j, \alpha | y(t-1))}_{via\ chain\ rule\ and\ Bayes\ rule}$$

$$= \underbrace{f(y_t | \Theta, c_t = i)}_{(1)\ or\ (2)} \underbrace{f(\Theta | y(t-1))}_{prior\ pdf\ of\ \Theta}$$

$$\times \underbrace{f(c_t = i | c_{t-1} = j, \alpha)}_{(5)} \underbrace{f(\alpha | y(t-1))}_{prior\ pdf\ of\ \alpha} \underbrace{f(c_{t-1} = j | y(t-1))}_{prior\ pointer\ pdf}.$$

Recursive formulas for estimation of c_t, Θ and α via (6) are obtained using the marginalization of (6) firstly over the parameters Θ and α. It results in the posterior pdf $f(c_t = i, c_{t-1} = j | y(t))$, which is joint for both c_t and c_{t-1}. Further the resulted joint pdf should be again marginalized over the values of c_{t-1} for obtaining the posterior pdf $f(c_t = i | y(t))$ of the current pointer.

3.1 Component Parameters

The integral of (6) over Θ is evaluated by substituting the point estimates of θ_i and r_i available from the previous time instant $t - 1$ and the currently measured y_t into the corresponding i-th normal component, either (1) or (2). The mentioned point estimates of parameters of the i-th component are computed based on using the conjugate prior Gauss-inverse-Wishart pdf with the recomputable (initially chosen) statistics $(V_{t-1})_i$ and $k_{i;t-1}$ in the Bayes rule, which according to [5, 16] gives the algebraic recursion for static components

$$(V_t)_i = (V_{t-1})_i + w_{i;t} \begin{bmatrix} y_t \\ 1 \end{bmatrix} [y_t, 1], \tag{7}$$

for dynamic components

$$(V_t)_i = (V_{t-1})_i + w_{i;t} \begin{bmatrix} y_t \\ \psi_{t-1} \end{bmatrix} [y_t, \psi_{t-1}], \tag{8}$$

and valid for both of them

$$\kappa_{i;t} = \kappa_{i;t-1} + w_{i;t},\tag{9}$$

where $w_{i;t}$ will be explained later. The needed point estimates are computed at time t for each component as follows [16]:

$$(\hat{\theta}_t)_i = V_1^{-1} V_y, \quad (\hat{r}_t)_i = \frac{V_{yy} - V_y' V_1^{-1} V_y}{\kappa_{i;t}},\tag{10}$$

where $(V_t)_i$ is partitioned (for simplicity with the omitted subscript i)

$$(V_t)_i = \begin{bmatrix} V_{yy} & V_y' \\ V_y & V_1 \end{bmatrix},\tag{11}$$

so that in the static case V_{yy} is the square matrix of the dimension N of the vector y_t, V_y' is N-dimensional column vector and V_1 is scalar. For dynamic components (2), the partition changes according to the memory length n used in the regression vector ψ_{t-1}, i.e., V_y and V_1 become matrices of appropriate dimensions. The substitution of (10) and y_t into the corresponding i-th normal pdf provides the proximity of each component to the current data item.

3.2 Pointer Parameters

Similarly, the integral of (6) over α provides the computation of its point estimate using the previous-time statistics denoted by ϑ_{t-1} of the conjugate prior Dirichlet pdf according to [4]. Here the mentioned statistics is the square m_c-dimensional matrix, whose entries for $c_t = i$ and $c_{t-1} = j$ are recursively computed in the following way:

$$\vartheta_{i|j;t} = \vartheta_{i|j;t-1} + W_{i,j;t},\tag{12}$$

where $W_{i,j;t}$ will be explained a bit later, and which was introduced by [17] with the approximation based on the Kerridge inaccuracy [25]. However, here, for simplicity, it is updated similarly to [5], but modified for the dynamic pointer model. The point estimate of α is then obtained by simple normalizing the updated statistics

$$\hat{\alpha}_{i|j;t} = \frac{\vartheta_{i|j;t}}{\sum_{k=1}^{m_c} \vartheta_{k|j;t}}.\tag{13}$$

3.3 Component Weights

Here the above denotations $w_{i;t}$ and $W_{i,j;t}$ are explained. After the described marginalization the posterior pdf $f(c_t = i, c_{t-1} = j | y(t))$ is obtained by entry-wise multiplying the proximity obtained from each component, the previous-time point estimate of α (13) and the prior pointer pdf $(c_{t-1} = j | y(t - 1))$. The last is the weight of the components at the previous time instant, and it is denoted by $w_{j;t-1}$ and expresses the (initially chosen and then actualized) probability of the activity of the j-th component at time $t - 1$.

For all $i, j \in \{1, 2, \ldots, m_c\}$, the posterior pdfs $f(c_t = i, c_{t-1} = j | y(t))$ are entries denotes by $W_{i,j;t}$ of the square m_c-dimensional matrix, which is normalized and summed up over rows to obtain the posterior pdf $f(c_t = i | y(t))$. The last provides the updated weight $w_{i;t}$ of each i-th component at time t. The maximal weight $w_{i;t}$ defines the currently active component, i.e., the point estimate of the pointer c_t at time t.

3.4 Initialization Problem Specification

The outlined relations are summarized as the following algorithm steps performed on-line for $t = 2, \ldots$:

1. Measure the new data y_t.
2. Obtain proximities of all components, using the previous-time point estimates (10).
3. Multiply entry wise the proximities, the prior weighting vector w_{t-1} and the previous-time point estimate $\hat{\alpha}_{t-1}$.
4. The result of this entry-wise multiplication is the matrix with entries $W_{i,j;t}$. Normalize this matrix.
5. Perform the summation of the normalized matrix over rows and obtain the updated vector w_t with entries $w_{i;t}$.
6. Update all statistics, using $w_{i;t}$ and $W_{i,j;t}$ according to (7) or (8), (9), and (12).
7. Recompute the point estimates of all parameters according to (10) and (13) and go to Step 1.

Thus, the initialization of this on-line part of the algorithm lies in setting at time $t = 1$:

- the number of components m_c,
- the initial statistics of all components $(V_0)_i$, $\kappa_{i;0}$ and the pointer statistics ϑ_0 (the initial estimates in Steps 2 and 3 are computed from them),
- the initial m_c-dimensional weighting vector w_0,

where m_c and $(V_0)_i$ are the key ones and they will be the focus of the subsequent sections. The rest of statistics can be initialized either uniformly or randomly in combination with their updating by prior data.

4 Expert-Based Mixture Initialization

The proposed initialization is based on convincing that in the beginning of the mixture estimation (as well as generally description of the multi-modal system) in a specific domain some type of prior or expert knowledge is always available. Such a kind of the knowledge can be in the form of specially previously measured data, realistic simulations (e.g., from Aimsun (www.aimsun.com) in the traffic flow control area) or, at least, the expert information about the expected number of components (disease symptoms in medicine, types of failures in car diagnostics, success in elections, etc.).

Anyway the start of the estimation is always critical due to a risk of dominance of a single active component resulted from the temporary non-activeness of others as well as noisy data. This can lead to joining other components and finally the failure of the estimation. To avoid the mentioned dominance the following expert-based procedures can be performed:

- fixing the covariance matrices of components as diagonal ones with entries 0.1 and running their estimation later, which is very simple and effective way;
- detection of the initial component centers by the visual analysis;
- repeated use of the prior data sample inspired by [4, 23].
- suppressing the influence of the first measured data on the estimation to support the initial estimates obtained from the initial statistics to produce proper weights of components based on [23].

These ways of processing the prior data to extract the information necessary for a successful initialization is described below. Thus, in this section the time instant t corresponds to prior data items. The implementation is prepared in the open source programming environment Scilab (www.scilab.org).

To determine the area of the interest in the data-parameter space it is suitable to work with the normalized data with zero expectations and the unit covariance matrices. This is reached by extracting the mean value from each prior data item and division by the standard deviation. However, it is not a necessary condition.

4.1 Static Component Initialization

For the initialization of static components (1) it is extremely important to detect the initial centers of clusters in the data space. This task covers both the determination of the number of components and of the initial statistics. Covariance matrices for the normalized data could be used as diagonal ones with entries 0.1.

For this aim the prior data sample is processed as follows. Individual entries of the multidimensional vector y_t are visualized by pairs against each other. The analysis of the visualization gives a possibility to distinguish the number of plotted components and get their centers. Here for demonstration, the real data sample measured on a driven vehicle is taken, where the vector y_t contains the following entries: (i) $y_{1;t}$ is the instantaneous fuel consumption [μl], (ii) $y_{2;t}$ is the vehicle speed [km/h], (iii) $y_{3;t}$ is pressing the gas pedal [%], (iv) $y_{4;t}$ is the engine speed [rpm]. The sampling period is 1 s. The number of prior data is 400.

Two-dimensional clusters of each variable are shown in Fig. 1. The visualization represents the upper triangular matrix of figures, where each row corresponds to the entry of the vector y_t from $y_{1;t}$ to $y_{4;t}$ plotted firstly against itself and then against the rest of entries. The normalized data with zero expectations and unit variances are used, which means that values on axes do not express real ranges of data items. Individual figures are denoted by numbers $l, k \in 1, \ldots, N$ corresponding to the entries indices. Under assumption that the processed data are of a multi-modal character, clusters are clearly visible. Here three clusters are seen, thus $m_c = 3$. For detection of initial centers of components, figures 1-2, 2-3 and 3-4 located above the diagonal are of the main interest.

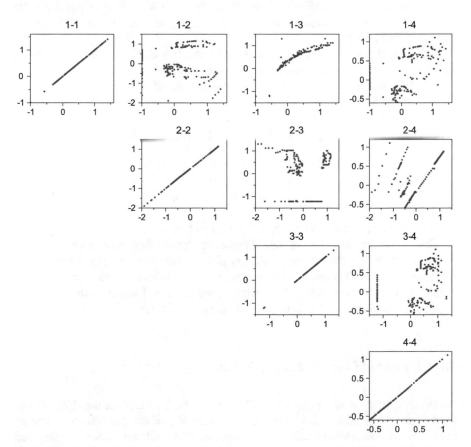

Fig. 1 Visualization of pairs of the normalized data vector against each other. Notice visible clusters plotted in figures denoted by 1-2, 2-3 and 3-4

Table 1 Initial centers of static components

Data entry	s_1	s_2	s_3
$y_{1;t}$	0	0.5	1
$y_{2;t}$	−0.5	1	−1
$y_{3;t}$	0.5	−1.2	0.5
$y_{4;t}$	−0.3	0.7	0

Figure 1-2 exhibits three clusters at positions $[0, -0.5]$, $[0.5, 1]$ and $[1, -1]$, which indicate three positions of clusters of the variable $y_{2;t}$: i.e., -0.5, 1 and -1. These values are explored in the second figure 2-3 on the x axis, where the variable $y_{2;t}$ is shown. Figure 2-3 gives the coordinates $[-0.5, 0.5]$, $[-1, -1.2]$ and $[1, 0.5]$, which provide positions 0.5 a -1.2 for the entry $y_{3;t}$. Using them in figure 3-4 the centers of components between entries $y_{3;t}$ and $y_{4;t}$ are detected as $[0.5, -0.3]$, $[0.5, 0.7]$ and $[-1.2, 0]$.

Based on this visual analysis, positions of the four-dimensional initial centers denoted by s_i $\forall i\{1, \ldots, m_c\}$ are summarized in Table 1.

The i-th initial center is substituted into the initial statistics $(V_0)_i$ of the i-component as follows:

$$
(V_0)_i = \begin{bmatrix}
1 & 0 & 0 & 0 & (s_1)_i \\
0 & 1 & 0 & 0 & (s_2)_i \\
0 & 0 & 1 & 0 & (s_3)_i \\
0 & 0 & 0 & 1 & (s_4)_i \\
(s_1)_i & (s_2)_i & (s_3)_i & (s_4)_i & 1
\end{bmatrix}
\tag{14}
$$

where $(s_l)_i$ $\forall l\{1, \ldots, N\}$ is the l-th entry of the vector s_i.

Thus, the expert-based initialization procedure for static components includes the steps: (i) normalize data (optionally); (ii) plot all data entries against each other, (iii) find subsequently positions of clusters in corresponding figures (here above the diagonal). The constructed initial statistics is used in the on-line part of the estimation algorithm. Validation of the approach is discussed in Sect. 5.

4.2 Dynamic Component Initialization

A character of dynamic components (2) requires both to support the dynamics of models and to avoid a preliminary dominance of any of components due to noisy data. The following initialization procedures can be considered (notice that they can be also combined).

4.2.1 Static Case Imitation

The first one is to imitate the static case described above and to detect both the number of components and their initial centers, using the visual analysis of prior data. The initial statistics $(V_0)_i$ in this case is constructed with the help of substituting a matrix of the form (14) with the detected initial centers instead of its part V_1 in (11). The rest of corresponding matrix entries are zero values. Such the initialization can be in many cases efficient, i.e., for data with the rather slow dynamics.

4.2.2 Initial Centers with Support of Dynamics

Another option is to combine the above approach with diagonal matrices, entries of which represent the chosen initial model dynamics. In this case (using the diagonal noise covariance matrix too) the component is decomposed into independent equations (in dependence on a dimension of the vector y_t). This allows to use the stabilized positions of centers for the initial statistics.

Construction of the initial statistics $(V_0)_i$ is based on the fact that the initial centers detected for static components are their constant expectations. Thus for the dynamic model (here for simplicity for the first order component with $n = 1$) the constant in (2), or precisely (4), is determined from

$$y_{l;t} - (a_{l|l})_i y_{l;t-1} - (s_l)_i, \tag{15}$$

where $\{a_i, s_i\} \in \theta_i$ of the i-th component, and $(a_{l|l})_i$ is the diagonal entry of the matrix of regression coefficients a_i with $l \in \{1, \ldots, N\}$, and $(s_l)_i$ is the entry of the vector s_i. Then the diagonal entry $(a_{l|l})_i$ expressing the dynamics (a small value about 0.1 brings more dynamics, and a value approaching 1 corresponds to slow dynamics) can be used for constructing the initial statistics. For the previous example, the initial statistics of the i-th strongly dynamic component is constructed by substituting

$$V_y = \begin{bmatrix} 0.1 & 0 & 0 & 0 \\ 0 & 0.1 & 0 & 0 \\ 0 & 0 & 0.1 & 0 \\ 0 & 0 & 0 & 0.1 \\ (s_1)_i & (s_2)_i & (s_3)_i & (s_4)_i \end{bmatrix} \tag{16}$$

into (11). It also defines V'_y, and the rest of submatrices are the unit matrices.

4.2.3 Repeated Use of the Data Sample

Another expert-based procedure, which is rather helpful in the initialization mostly under condition of the lack of data is the repeated use of the available prior data sample [4, 23]. Firstly the estimation starts according to the algorithm from Sect. 3.4

with small diagonal initial statistics $(V_0)_i$. The actualized statistics after the course of the estimation with the whole sample of prior data are used as the new initial one, and the estimation algorithm starts again. The resulted updated statistics are used as initial for the on-line estimation.

This way of initialization can be also combined with weighting the initial statistics to suppress the influence of data in the beginning of the algorithm running, which is described below.

4.2.4 Weighting the Initial Statistics

This initialization approach is primarily based on [23], which in the considered context takes the following form.

The prior or expert given data are firstly substituted in the extended regression vectors $\left[y_t, \ \psi_{t-1}\right]'$ used then in the statistics update (8). The amount of the used extended regression vectors should correspond to the number of parameters (regression coefficients) to be estimated. The statistics $\kappa_{i;0}$ expresses the number of the used data.

The Bayesian estimation is strengthened with gradually measuring new data, which means that a weight of the new data item is inversely proportional to the statistics, and therefore the possible disturbance in data takes the inversely proportional effect on the estimation. Thus the same prior regression vectors multiplied by the chosen weight are used again for the initial statistics as follows:

$$(V_0)_i = \mu(V_0)_i, \quad \kappa_{i;0} = \mu, \tag{17}$$

where μ expresses the number of the used prior data items, which means that the first newly measured data item takes the effect $\frac{1}{\mu}$ on the estimation.

Improvements brought by the mentioned initialization methods appear primarily in the speed of finding the stabilized estimates of regression coefficients during the on-line mixture estimation. Validation of the enumerated approaches is presented below.

5 Experiments

The initialization of the mixture estimation algorithm can be validated in accordance with the following criteria.

5.1 Weight Evolution

The first one concerns with the initialized number of components, which is identical both for the static and dynamic components. It is verified by the evolution of the component weights during the on-line part of the estimation algorithm using 6400 data. For better visibility, fragments with 1200 data items are shown. The evolution should demonstrate a reasonable way of switching the components. In that case it confirms that the model is correctly established. For the prior data used in Sect. 4.1 the evolution of the corresponding entries of the weighting vector w_t of three detected components is shown in Fig. 2.

It can be seen that (i) the components switch in a reasonable way, (ii) the plotted values of probabilities are mostly approaching 1 or 0, which means the unambiguous decision for the currently active component.

5.2 Parameter Estimate Evolution

The evolution of the component centers for static components (1) and of the estimates of regression coefficients for dynamic ones (2) is a sufficient indicator of the successful initialization.

Stabilization of positions of component centers in the data space after their initial search indicates that the estimation is correct. Otherwise, if some resulting centers are identical or very close one to another, this mostly signalizes that too many components are chosen, and their number should be reduced.

The evolution of the component centers can be seen in Fig. 3 in the parameter space for the normalized entries $y_{1;t}$ and $y_{2;t}$, and $y_{3;t}$ and $y_{4;t}$ plotted against each other.

The evolution of the estimation of regression coefficients of the components is shown in Fig. 4. The stabilization of the estimation can be seen, where after the initial search the steady-state is reached.

5.3 Validation via Data Prediction

The graphically represented comparison of the predicted data items obtained from components with the substituted estimates and the real data is shown in Fig. 5. Normalized entries of the data vector y_t are shown. Graphs demonstrate the coincidence between predictions and real data items.

The presented results are shown for the combination of the visual analysis with the dynamics support initialization, the repeated use of the prior sample and weighting the initial statistics, which gives the minimal prediction error in comparison with other combinations.

Fig. 2 The evolution of the
activity of three components.
Notice that values of the
weights are approaching 0
or 1

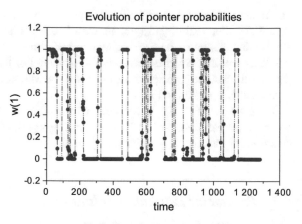

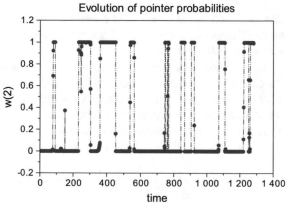

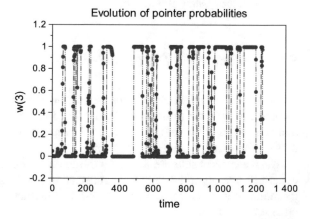

Fig. 3 Evolution of three component centers. The start position is denoted by 'x', and the end of the search is marked by 'o'. The density of points corresponds to the speed of movement

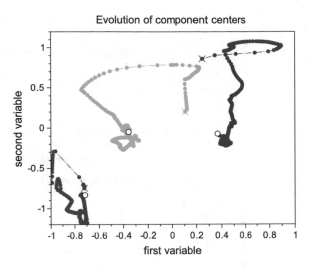

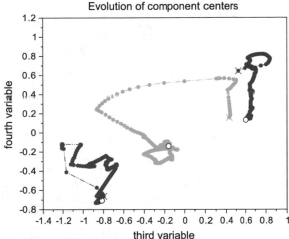

5.4 Closely Located Components

Another series of experiments has been performed with a set of data with the same sampling period, where the vector y_t contains three entries: (i) $y_{1;t}$ is position of the gas pedal [%], (ii) $y_{2;t}$ is lateral acceleration in multiples of gravimetric acceleration, and (iii) $y_{3;t}$ is road altitude (height above sea level) [m]. The number of prior data was 200.

As it can be seen in Fig. 6, it is difficult to distinguish individual components with the help of the prior data visualization. The plot 2–3 can be used for a prior guess about 5 components with coordinates [2, −0.5], [0, −1], [1, 1.2], [−1, 1.5],

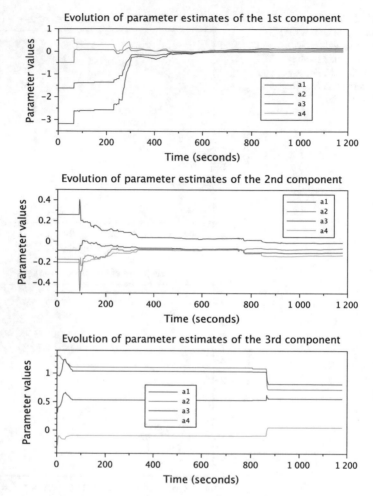

Fig. 4 Evolution of regression coefficients of three components. Notice that after the initial search the stabilized state is reached. In the bottom figure the initialization has given the already stabilized values

$[-3, 1]$. However, the rest of plots does not confirm that, and centers of components cannot be constructed.

Under assumption of existing 5 components the prior data set with 200 values of the gas pedal position can be used for discretization within 5 intervals, which are further used as prior known values of the pointer. Another possibility, often available in practice, is a help of an expert, who can assign values of modeled entries to components. Here the technique of the repeated use of the data sample [4, 23] described in Sect. 4.2.3 is combined with application of 5 discretized values of the gas pedal position as the known pointer. The estimation is running firstly for 200 prior data items with small diagonal initial statistics of 5 components. Only the statistics of

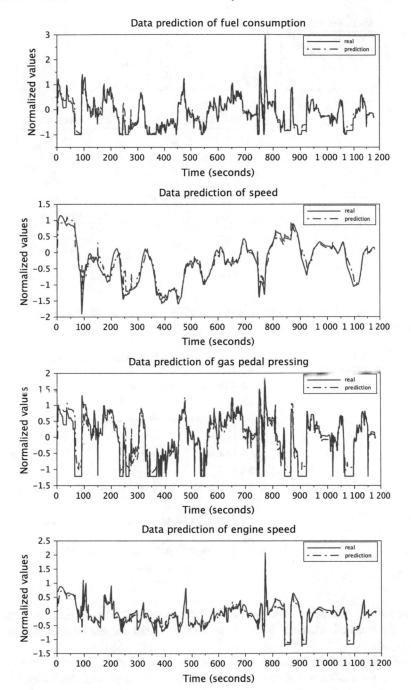

Fig. 5 Results of data prediction. Notice that predicted values correspond to real data items

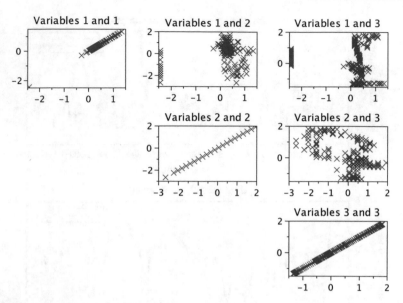

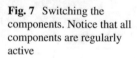

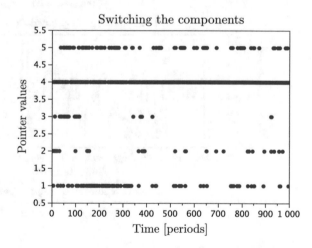

Fig. 6 Visualization of data entries against each other. Here only plot 2–3 can be used for the initialization purposes

Fig. 7 Switching the components. Notice that all components are regularly active

the active component (according to the pointer supposed to be known) are updated. Then the resulted updated statistics are used as initial for the on-line part of the algorithm.

Switching the components is shown in Fig. 7, where the point estimates of the pointer as the maximum entries of the weighting vector are plotted. All 5 components demonstrate activity, which means that the choice of the number of components is adequate.

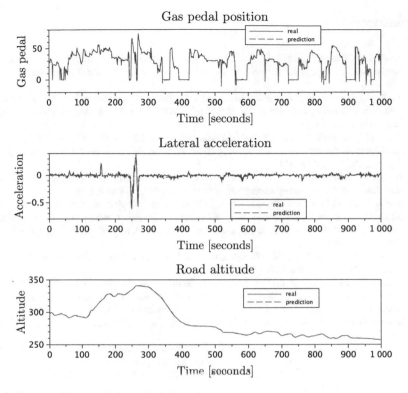

Fig. 8 Results of data prediction with 5 dynamic components

Dynamic components of the first order are used. Data prediction of modeled entries of the vector y_t can be found in Fig. 8.

6 Conclusion

The presented approach is based on the availability of prior or expert data, which is always the case in real application fields. Thus the intervention of an expert in processing the prior data is realistic and, as it can be seen, advantageous for such a critical task as the mixture initialization. This paper focuses on initialization of mixtures of normal components. However, the present research project aims at the recursive estimation of mixtures of different distributions (namely, categorical, exponential, uniform components), which all require specific initialization approaches. This will be part of the future project work.

Acknowledgements The research was supported by project GAČR GA15-03564S.

References

1. Park, B.-J., Zhang, Y., Lord, D. (2010). Bayesian mixture modeling approach to account for heterogeneity in speed data, *Transportation Research Part B: Methodological*, vol. 44, 5, pp. 662–673.
2. Yoshigoe, K., Dai, W., Abramson, M., Jacobs, A. (2015). Overcoming invasion of privacy in smart home environment with synthetic packet injection, In: *TRON Symposium (TRONSHOW)*, Tokyo, Japan, 2014, pp. 1–7.
3. Yu, J. (2011). Fault detection using principal components-based Gaussian mixture model for semiconductor manufacturing processes, *IEEE Transactions on Semiconductor Manufacturing*, vol. 24, 3, pp. 432–444.
4. Kárný, M., Böhm, J., Guy, T. V., Jirsa, L., Nagy, I., Nedoma, P., Tesař, L. (2006). Optimized Bayesian dynamic advising: theory and algorithms, Springer-Verlag London.
5. Kárný, M., Kadlec, J., Sutanto, E. L. (1998). Quasi-Bayes estimation applied to normal mixture, in: *Preprints of the 3rd European IEEE Workshop on Computer-Intensive Methods in Control and Data Processing (eds. J. Rojíček, M. Valečková, M. Kárný, K. Warwick)*, CMP'98 */3./*, Prague, CZ, pp. 77–82.
6. Gupta, M. R., Chen, Y. (2011). Theory and use of the EM method, in: *Foundations and Trends in Signal Processing*, vol. 4, 3, pp. 223–296.
7. Boldea, O., Magnus, J. R. (2009). Maximum likelihood estimation of the multivariate normal mixture model, *Journal Of The American Statistical Association*, vol. 104, 488, pp. 1539–1549.
8. Wang, H. X., Luo, B., Zhang, Q. B., Wei, S. (2004). Estimation for the number of components in a mixture model using stepwise split-and-merge EM algorithm, *Pattern Recognition Letters*, vol. 25, 16, pp. 1799–1809.
9. McGrory, C. A., Titterington, D. M. (2009). Variational Bayesian analysis for hidden Markov models, *Australian & New Zealand Journal of Statistics*, vol. 51, pp. 227–244.
10. Šmídl, V., Quinn, A. (2006). The Variational Bayes method in signal processing, Springer-Verlag Berlin Heidelberg.
11. Frühwirth-Schnatter, S. (2006). Finite mixture and Markov switching models, Springer-Verlag New York, 2006.
12. Doucet, A., Andrieu, C. (2001). Iterative algorithms for state estimation of jump Markov linear systems, *IEEE Transactions on Signal Processing*, vol. 49, 6, pp. 1216–1227.
13. Chen, R., Liu, J. S. (2000). Mixture Kalman filters, *Journal of the Royal Statistical Society: Series B (Statistical Methodology)*, vol. 62, pp. 493–508.
14. Aggarwal, C. Ch. (2015). Outlier Analysis. In: *Data Mining: The Textbook*, Springer International Publishing, pp. 237–263.
15. Fukunaga, K. (2013). Introduction to Statistical Pattern Recognition, series Computer science and scientific computing, Elsevier Science.
16. Peterka, V. (1981). Bayesian system identification. In: *Trends and Progress in System Identification (ed. P. Eykhoff)*, Oxford, Pergamon Press, 1981, pp. 239–304.
17. Nagy, I., Suzdaleva, E., Kárný, M., Mlynářová, T. (2011). Bayesian estimation of dynamic finite mixtures, *Int. Journal of Adaptive Control and Signal Processing*, vol. 25, 9, pp. 765–787.
18. Melnykov, V., Melnykov, I. (2012). Initializing the EM algorithm in Gaussian mixture models with an unknown number of components, *Computational Statistics & Data Analysis*, vol. 56, 6, pp. 1381–1395.
19. Kwedlo, W. (2013). A new method for random initialization of the EM algorithm for multivariate Gaussian mixture learning, in: *Proceedings of the 8th International Conference on Computer Recognition Systems CORES 2013*, (eds. R. Burduk, K. Jackowski, M. Kurzynski, M. Wozniak, A. Zolnierek), Springer International Publishing, Heidelberg, pp. 81–90.
20. Blömer, J., Bujna, K. (2013). Simple methods for initializing the EM algorithm for Gaussian mixture models, *CoRR*, vol. abs/1312.5946, arxiv.org/abs/1312.5946.
21. Ning, H., Yuxiao Hu, Y., Huang, T. (2008). Efficient initialization of mixtures of experts for human pose estimation, in: *Proceedings of the IEEE International Conference on Image Processing*.

22. Paclik, P., Novovičová, J. (2001). Number of components and initialization in Gaussian mixture model for pattern recognition, in: *Proceedings of the 5th International Conference on Artificial Neural Networks and Genetic Algorithms, ICANNGA 2001*, Prague, Czech Republic, pp. 406–409.
23. Kárný, M., Nedoma, P., Khailova, N., Pavelková, L. (2003). Prior information in structure estimation, *IEE Proceedings, Control Theory and Applications*, vol. 150, 6, pp. 643–653.
24. Suzdaleva, E., Nagy, I., Mlynářová, T. Expert-based initialization of recursive mixture estimation. In Proceedings of 2016 IEEE 8th International Conference on Intelligent Systems IS'2016, p. 308–315, Sofia, Bulgaria, September 4–6, 2016.
25. Kerridge, D. (1961). Inaccuracy and inference, *Journal of Royal Statistical Society B*, vol. 23, pp. 284–294.

Spatiotemporal Parameter Estimation of Thermal Treatment Process via Initial Condition Reconstruction Using Neural Networks

M Hadjiski, Nencho Deliiski and Aleksandra Grancharova

Abstract In this paper the design of control systems of periodical thermal treatment processes (TTP) with distributed parameters modeled by partial differential equations (PDEs) is considered. The main problem to decide is the estimation of the initial charge parameters—size, humidity, temperature and the relative load, which are all immeasurable. The investigation is based on first-principle models of the internal and external heat-exchange. Initially, after deriving the PDEs is created a representative TTP set by simulation using relevant combinations of charging parameters. To obtain their real estimates, the only measurable heating medium temperature, informative only during the first TTP stage, is applied. A cluster of N-nearest neighborhoods is found around the charge experimental temperature curve. A local situation-based dynamic neural network is learned to assess the charging parameters. They are implemented to define the optimal heating time of the current charge using another static neural network. Finally some aspects of industrial application of the proposed approaches are discussed.

Keywords Distributed parameter systems (DPS)
Partial differential equation (PDE) · Neural network (NN)
Thermal treatment process (TTP) · Parameter estimation

M Hadjiski (✉)
Institute of Information and Communication Technologies, BAS, Acad. G. Bonchev str. Bl. 2, 1113 Sofia, Bulgaria
e-mail: hadjiski@uctm.edu; zdravkah@abv.bg

M Hadjiski · A. Grancharova
Department of Industrial Automation, University of Chemical Technology and Metallurgy, St. Kliment Ohridski blvd. 8, 1796 Sofia, Bulgaria
e-mail: alexandra.grancharova@abv.bg

N. Deliiski
Faculty of Forest Industry, University of Forestry, St. Kliment Ohridski blvd. 10, 1796 Sofia, Bulgaria
e-mail: deliiski@netbg.com

© Springer International Publishing AG, part of Springer Nature 2019
M Hadjiski and K T Atanassov (eds.), *Intuitionistic Fuzziness and Other Intelligent Theories and Their Applications*, Studies in Computational Intelligence 757,
https://doi.org/10.1007/978-3-319-78931-6_4

1 Introduction

Wood traditionally continues to be important constructive material in furniture and also in the similar industries. With the increased competition in wood-making industry, the realized heat energy consumption, the growing capacity and fulfilling the strict tight technological requirements has become a central objective in management, control and operation of the one of the most heat- and time-consuming processes in transforming the raw material into an industrial product—the wood thermal treatment process (TTP).

Thermal treatment of wood materials comprises spatially distributed technological processes. They are described by partial differential equations (PDE). Various approaches have been developed during the last half century in order to overcome the numerous difficulties in modeling, simulation and control arising from the comparison with the lumped parameter systems Lions [1], Butkovskii [2], Ray and Lainiotis [3]. The academic-oriented research created principally new directions in transforming the infinite space presentation by finite dimension modeling in the linear case Curtain and Zwart [4], Morris [5], Christofides [6]. Considerable research interest was focused on model-reduction-based techniques Zheng and Hoo [7], Xie et al. [8]. The investigations have been enhanced towards nonlinear distributed parameter systems modeling and control Dufour et al. [9], Shvartsman et al. [10], Xie et al. [8], Balsa-Canto et al. [11]. Model-based approaches of DPS achieved significant success by using deeply studied techniques of Model Predictive Control Brosilow and Joseph [12], Dubljevic and Christofides [13], Aggelogiannaki et al. [14], Dufour et al. [9], Zheng and Hoo [7].

In parallel with the general model-based direction of optimization and control of distributed parameter systems, an observer-based stream of successful investigations have been accomplished Hidayat et al. [15]. A variety of techniques was extensively studied: finite-dimensional adaptive observers Lilly [16], Curtain et al. [17], backstepping observers Smyshlyaev and Krstic [18], sliding-mode observer Miranda et al. [19], exponential observer Garcia et al. [20], state reconstruction based on partial measurements Garcia et al. [21].

During the last decade several intelligent techniques have been incorporated in modeling and model-predictive control of distributed parameter systems Hadjiski and Deliiski [22], and especially the implementation of artificial neural networks (NN) Aggelogiannaki and Sarimveis [23], Aggelogiannaki et al. [14]. Though that wood thermal treatment processes were deeply studied in different technological aspects Shubin [24], Videlov [25], Hill [26], Esteves and Pereira [27], Pervan [28], only a few investigations are addressed to modeling and control of TTP using the listed above achievements in the area of distributed parameter systems. Available results are formed mainly in modeling Khattabi and Steinhagen [29–31], Moreno and Devlieger [32], Younsi et al. [33, 34].

In this work the earlier developed system for 2D and 3D modeling of frozen and not frozen logs and prismatic wood materials Deliiski [35–39], Deliiski and Dzurenda [40], as well the proposed systems for advanced TTP control Hadjiski and

Deliiski [22, 41] are enhanced with emphasis on reducing the uncertainty of initial immeasurable charging conditions which are critical for successful model-based estimation, control and optimization.

2 Thermal Treatments Processing (TTP)

2.1 TTP Principles

The thermal treatment is a periodical process of wood material heating in order to reach a given average mass temperature of the charge subjected to prescribed requirements according to the admissible surface temperature and the internal temperature gradients. TTP is carried out in autoclaves with steam as a heating agent or in pits using hot water. The cross-section of a typical autoclave is presented in Fig. 1.

The thermal treatment is a spatiotemporal nonlinear process. Typical temperature profiles for some characteristic points of the cross-section of beech logs with an initial temperature 0 °C and −10 °C during their steaming in an autoclave with diameter of 2.4 mm and length of 9.0 m, obtained by computer simulation Deliiski and Dzurenda [40] are presented in Fig. 2 (up) and Fig. 2 (down), respectively.

Two distinct heating processes can be observed in Fig. 2: (*i*) fast for the external heating medium temperature θ_m and (*ii*) slow for the internal points of the wood space.

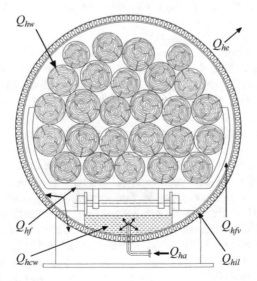

Fig. 1 Autoclave cross-section

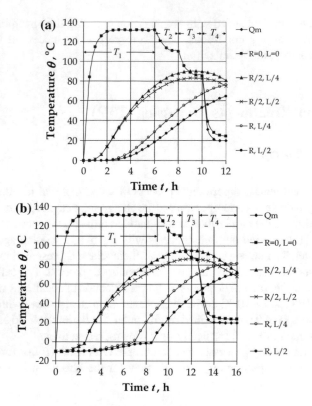

Fig. 2 Temperature time profiles during TTP for not frozen (a) and frozen (b) beech logs

The only temperature of the heating agent θ_m and the steam flow rate F_m can be measured on-line. Due this strong drawback the only model-based approaches are suitable for control systems design, state estimation, dispatching operations, and production scheduling. Unfortunately the lack of measurements provokes strong difficulties in the TTP management.

The operational conditions of each charge can be different because of the considerable variations of the initial conditions—wood specie, size, moisture content, temperature of the wood and aggregate state of the water in it at the beginning of TTP, and the relative loading of the autoclave as well. At the same time some of these initial charging parameters are immeasurable. Unfortunately they determine the parameters, the initial and boundary conditions necessary to solve the first-principle models based on partial differential equations (PDE) Moreno and Devlieger [32], Deliiski [35–39], Deliiski and Dzurenda [40], Hadjiski and Deliiski [41].

The attempt to use as model parameters some 'average' or operator's given values causes models fail and discredit all efforts for advanced control implementation in real industrial conditions.

2.2 TTP Peculiarities from a Control Perspective Point of View

- Only θ_m and F_m are on-line available as a secondary (indirect measurements);
- The values of θ_m and F_m are available only less than 20% of the whole TTP duration;
- The initial changing conditions are immeasurable;
- The accuracy of the PDE-based model considerably depends on the initial changing conditions;
- The TTP is strongly nonlinear distributed parameter system, especially in dependence of the frozen/unfrozen charged timber;
- The mathematical TTP model is highly time-consuming due to the nonlinearities and the space distribution. Thus adopting the on-line calculations as a part of the control algorithm cannot be a relevant strategy. More appropriate would be:

  - To develop models with reduced dimensionality,
  - The biggest part of the calculations to be carried out off-line,
  - To accept a strategy of suboptimal control,

- The TTP is imposed on numerous constraints like Hadjiski and Deliiski [22]:

$$\frac{\partial \theta(x, y, z, t)}{\partial t} < \Gamma_1,$$
$$\frac{\partial \theta(x, y, z, t)}{\partial n} < \Gamma_2,$$
$$\theta(s) < \theta^{\max}, \tag{1}$$

where $\theta(x, y, z, t)$ is the wood temperature in the point with coordinates (x, y, z), t—time, n—the normal vector, $\theta(s)$—the surface temperature, s—the surface coordinates $\Gamma_1, \Gamma_2, \theta^{\max}$—the technological requirements;

- The TTP control system constraints depend strongly on the model parameters;
- In addition to the uncertainty due to the immeasurable initial charging parameters, the TTP is dependent on some specific bio- and morphological properties of the timber. They are immeasurable and stochastic Shubin [24], Moreno and Devlieger [32], Videlov [25].

3 Problem Statement

This work is a continuation of our previous investigations Deliiski [35], Hadjiski and Deliiski [41], Hadjiski and Deliiski [22], Deliiski and Dzurenda [40] via an extension of the functionality of the control system in order to improve the TTP performance. The research is focused towards the following below trends:

- Creation of a subsystem to estimate the real current parameters of the developed earlier mathematical models Deliiski [36, 38], Deliiski and Dzurenda [40] with an emphasis on the initial conditions' reconstruction which are critical for the model parameter formation.
- Estimation procedure fulfilled using a variety of hybrid technologies—mathematical modeling, simulation, statistics, neural networks, queries, learning.
- The available information about the input to the estimation procedure is scare, incomplete, imperfect and contradictory, namely:

  - The measurable temperature of the heating fluid θ_m;
  - The estimations of the human operator;
  - The data from the previous charges, received by the same estimation subsystem.

- Complete the developed already mathematical models mainly based on solving PDEs in order to describe the internal heat and mass transfer Deliiski [36, 38], Deliiski and Dzurenda [40] with a new additional model for the internal heat exchange description in order to determine the heating fluid temperature θ_m more accurately.
- Change the control strategy in the aspect of increasing the ratio off-line/on-line calculations in order to reduce the computation load by:

  - Creation of bases with a large number of situation-based time-profiles for wood-logs TTP considered as DPS by varying the model parameters, the initial and boundary conditions;
  - Reduce long iterative procedures;
  - Implement mainly parameterized representation of DPS behavior using neural networks.

- Increase the robust performance of the integrated control system by reducing the system uncertainty using more accurate model parameters.
- Broad implementation of learning procedures for neural network tuning, 'batch-to-batch' parameters corrections, improved internal weight constants.

Different kinds of constraints must be taken into account in the research:

- Technological constraints presented by the relations (1);
- The control system constraints

$$u_0 < u_0^{\max};$$

$$\sigma_P \leq \sigma_P^{\Gamma}.$$

- Computational load constraints according to the accepted industrial automation designed systems.

Several methods for observation-based state-space reconstruction for DPS are available in the literature e.g. see Hidayat et al. [15], Curtain et al. [17], Lilly [16], Smyshlyaev and Krstic [18].

In this investigation such approach was out of consideration because:

- In the case with TTP we have a measurement only for the heating medium temperature θ_m.
- We developed a verified and applied in industry mathematical model, based on the first principle.
- The nonlinearities in the process model are very complex and they do not correspond to the usual assumptions for smoothness when some methods of dimensionality reduction are used by DPS observers.
- The DPS observers have a number of challenges as the computational time, the sensitivity to bad data and outliers.

4 Mathematical Modeling of TTP

4.1 Mathematical Model for the Non-stationary Heating of Wood Materials During Their TTP

A basic task in heat modeling, developing and managing the technologies and control of TTP is the determination of the temperature in certain points from the volume of the wood materials, at any moment from the process of their heating and further conditioning in aerial medium.

During the heating of the wood materials along with the purely thermal processes, a mass-exchange occurs between the heating medium and the wood. The values of the moisture diffusion of the different wood species cross sectional to their fibers are hundreds of times smaller than the values of their temperature conductivity. In a longitudinal to the fibers direction, the temperature conductivity exceeds the moisture diffusion by more than a hundred times.

These facts determine not so big change in the content of water in the materials during their TTP, which lags significantly from the distribution of heat in them. This allows during the creation of a mathematical model to disregard the exchange of mass between the wood and the heating medium and the change in temperature in the materials to be viewed as a result of a purely thermo-exchange process, where the heat in them is distributed only through thermo-conductivity.

The process of heating the subjected to TTP wood materials with prismatic form can be modeled with the help of the following system of equations

$$c(\Theta, w, w_{fsp}) \cdot \rho(\rho_b, w)\frac{\partial \Theta(x, y, z, t)}{\partial t} = \frac{\partial}{\partial x}\left[\lambda_x(\Theta, w, w_{fsp}, \rho_b)\frac{\partial \Theta(x, y, z, t)}{\partial x}\right] +$$
$$\frac{\partial}{\partial y}\left[\lambda_y(\Theta, w, w_{fsp}, \rho_b)\frac{\partial \Theta(x, y, z, t)}{\partial y}\right] + \frac{\partial}{\partial z}\left[\lambda_z(\Theta, w, w_{fsp}, \rho_b)\frac{\partial \Theta(x, y, z, t)}{\partial z}\right] \quad (2)$$

with an initial condition:

$$\Theta(x, y, z, 0) = \Theta_0 \tag{3}$$

and the following boundary conditions:

- during TTP:

$$\Theta(0, y, z, t) = \Theta(x, 0, z, t) = \Theta(x, y, 0, t) = \Theta_m(t), \tag{4}$$

- during the conditioning in aerial medium of the heated materials:

$$\frac{\partial \Theta(0, y, z, t)}{\partial x} = -\frac{\alpha_{sx}(0, y, z, t)}{\lambda_{sx}(0, y, z, t)}[\Theta(0, y, z, t) - T_a], \tag{5}$$

$$\frac{\partial \Theta(x, 0, z, t)}{\partial y} = -\frac{\alpha_{sy}(x, 0, z, t)}{\lambda_{sy}(x, 0, z, t)}[\Theta(x, 0, z, t) - T_a], \tag{6}$$

$$\frac{\partial \Theta(x, y, 0, t)}{\partial z} = -\frac{\alpha_{sz}(x, y, 0, t)}{\lambda_{sz}(x, y, 0, t)}[\Theta(x, y, 0, t) - T_a], \tag{7}$$

where Θ, Θ_0, Θ_m, Θ_a, are the temperature, temperature of the wood at the beginning of TTP, temperature of the processing medium during TTP, and temperature of the aerial medium near the subjected to conditioning heated wood materials respectively, K. The meaning of the other notation is as follows:

c—specific heat capacity of the wood, J.kg^{-1}.K^{-1};

w—moisture content of the wood, kg.kg^{-1};

w_{fsp}—fiber saturation point of the wood specie, kg.kg^{-1};

λ_x, λ_y, and λ_z—thermal conductivities of the wood in radial, tangential and longitudinal anatomical directions respectively, W.m^{-1}.K^{-1};

λ_{sx}, λ_{sy}, and λ_{sz}—thermal conductivities on the surfaces of the wood materials in radial, tangential and longitudinal directions respectively, W.m^{-1}.K^{-1};

ρ—density of the wood, kg.m^{-3};

ρ_b—basic density of the wood, equal to the dry mass divided by green volume, kg·m^{-3};

α_{sx}, α_{sy} and α_{sz}—heat transfer coefficients between the respective surfaces (perpendicular to the coordinate axes x, y, z) of the subjected to TTP wood materials and the surrounding aerial medium, W.m^{-2}.K^{-1};

x—linear coordinate of each point from the thickness of the wood prism, m;

y—linear coordinate of each point from the width of the wood prism, m;

z—linear coordinate of each point from the length of the wood prism, m;

t—time, s.

For the solution of the created mathematical model a software package has been developed in the computing medium of Visual Fortran Professional. Through simulation of computer experiments this package allows to investigate the distribution of the temperature field in the volume of wood materials at different initial and boundary conditions and to form scientifically based technologies and control for TTP both in autoclaves and equipments, working under atmospheric pressure.

4.2   Modeling of the Energy Consumption of Autoclaves

The heat energy Q_{ha}, which is fed into the autoclave by the introduced in it water steam is consumed for:

- warming up of the subjected to steaming wood materials (Q_{hw});
- heating of the body of the autoclave and of the situated in it metal trolleys for positioning of the wood materials (Q_{hf});
- warming up of the heat insulating layer of the autoclave (Q_{hil});
- covering of the heat emission from the autoclave in the surrounding aerial space (Q_{he});
- filling in with steam the free (unoccupied by wood materials) part of the working volume of the autoclave (Q_{hfv});
- accumulating heat in the gathered in the lower part of the autoclave condenses water (Q_{hcw}).

On the grounds of the performed analyses, the shown on Fig. 1 structural model of distribution of the heat in the autoclaves for steaming of wood materials is suggested. By using this model, a mathematical description of the heat energy consumption and its equivalent of water steam are made, as well as a mathematical description of the heat and its corresponding steam balance of the autoclaves.

The total specific heat energy needed for TTP of 1 m^3 wood materials in autoclave for any moment $n{\cdot}\Delta\tau$ of TTP, q_{ha}^n, is equal to (in kWh·m^{-3}):

$$q_{ha}^n = q_{in}^n + q_{out}^n, \tag{8}$$

where $\Delta\tau$ is the step along the time of TTP in s, n is time level of TTP: $n = 0, 1, 2,..., N=\tau*/\Delta\tau$ and:

$$q_{in}^n = q_{hw}^n = q(P_1, P_2, P_3, P_4). \tag{9}$$

$$q_{out}^n = q_{hf}^n + q_{hil}^n + q_{he}^n + q_{hfv}^n + q_{hcw}^n = q(P_1, P_2, P_3, P_4, A). \tag{10}$$

The total specific heat flux $\frac{dq_{ha}^n}{d\tau}$, which provides the energy q_{ha}^n for any moment $n{\cdot}\Delta\tau$ of TTP can be determined (in kW) according to the following equation:

$$\frac{dq_{ha}^n}{d\tau} \approx \frac{3600 q_{ha}^n}{\Delta\tau}, \tag{11}$$

For the realization of automatic control of TTP with limited power of the heat generator, q_{source} (in kW), the following problem occurs: depending on the present limited heat power it is needed to determine the real law of increase in the temperature of the processing medium Θ_m during the initial part of TTP. For the solution of such problem, an algorithm for the calculation of the change in Θ_m after submission of the limited heat power to autoclave at the beginning of TTP until reaching of the technological acceptable maximal value of Θ_m^{max} is needed.

It is known that at the beginning of TTP the temperature Θ_m increases to curvilinear dependence. The separate sections of this dependence can be approximated by a part of exponent with respective time constant t_e for each step $\Delta\tau$ of TTP. This allows describing the increase of Θ_m during the initial part of TTP by the following equation:

$$\Theta_m^n = \Theta_m^{max} - \left(\Theta_m^{max} - \Theta_{m0}\right) \exp\left(-\frac{t}{t_e}\right), \qquad (12)$$

where Θ_m^n is the current value of the processing medium temperature in the autoclave for each moment $n \cdot \Delta\tau$ of TTP, K, and:

Θ_{m0}—initial medium temperature in the autoclave, K;
Θ_m^{max}—maximal technologically allowable medium temperature in the autoclave, K;
t—current time of TTP, equal to $n \cdot \Delta\tau$, s;
t_e—time constant of the exponential increase of separate sections of Θ_m during the initial part of TTP, s.

For determination of t_e and computation of Θ_m^n for each moment $n \cdot \Delta\tau$ of the initial part of TTP the method for optimization with variable reverse step is suitable for use. For this purpose the following optimization criteria can be used:

$$q_{source} - \partial_h \leq \frac{dq_{ha}^n}{dt} \leq q_{source} + \partial_h, \qquad (13)$$

where by ∂_h (in kW) the setting of the limits of localization of $\frac{dq_{ha}^n}{dt}$ is carried out.

For the determination of the time constant t_e in Eq. (12), an algorithm and a subroutine to the software package were created. Depending on the limited power of the heat generator, q_{source}, they realize the optimization procedures for the calculation of Θ_m^n. On Fig. 3 the block-scheme of the suggested algorithm is shown.

In the main program of the software package the conditions of Eq. (13) is checked.

When this condition is satisfied, the subroutine is ignored and the change in Θ_m according to Eq. (12) for the next step $\Delta\tau$ is calculated. When the condition (13) is not satisfied, the execution of the subroutine begins by checking the inequality $\frac{dq_{ha}^n}{dt} > q_{source} + \partial_h$.

If this inequality is satisfied, it means that the value of Θ_m^n, which for the calculation of the total heat energy q_{ha}^n and of the current value of $\frac{dq_{ha}^n}{dt}$ had been used, is larger than the value of Θ_m^n, by which the condition (12) will be satisfied. In this case a procedure

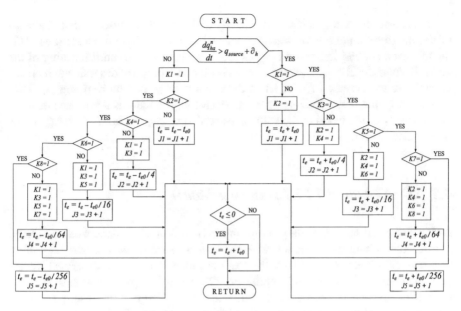

Fig. 3 Block-scheme of the algorithm for determination of t_e in Eq. (12)

of increasing of t_e (by consistently adding of 1/4, 1/16, 1/64, and 1/256 part from the initial value t_{e0} to the current value of t_e—according to the recommendations of the method for optimization with variable reverse step) starts, which provides a decrease of Θ_m^n until reaching of such a value, by which the condition (13) is satisfied.

If the condition $\frac{dq_{ha}^n}{dt} > q_{source} + \partial_h$ is not satisfied, it means that the value of Θ_m^n, which for the calculation of the current values of q_{ha}^n and $\frac{dq_{ha}^n}{dt}$ had been used, is less than the value of Θ_m^n, by which the condition (13) will be satisfied. In this case a procedure of decreasing of t_e (by consistently subtracting from the current t_e value of 1/4, 1/16, 1/64, and 1/256 part of the initial value t_{e0}) starts, which provides an increase of Θ_m^n until reaching of such a value, by which the condition (13) is satisfied.

When the condition (13) is accomplished that means that during the next step $\Delta\tau$ of TTP the energy $\Delta\tau \cdot q_{ha}$, which with the determined true value of Θ_m^n is calculated will be equal to the energy $\Delta\tau \cdot q_{source}$.

The main consumer of the heat energy in TTP are the wood materials in the autoclave. That is why during the optimization procedure for the determination of t_e the total energy of the autoclave $\Delta\tau \cdot q_{ha}$ is most impacted by the change of the heat consumption of the wood $\Delta\tau \cdot q_{hw}$.

As the thickness of the wood materials d increases, the heat in the wood materials is distributed slower. This means that with an increase of d, a higher increase in Θ_m^n during any next step $\Delta\tau$ until warming up of the surface layers of the wood materials and reaching of the equality $\Delta\tau \cdot q_{ha} = \Delta\tau \cdot q_{source}$ is needed. This provides for a faster increase in Θ_m^n with an increase of d during the initial part of TTP.

It is known that the specific heat capacity of the wood increases with an increase of the moisture content w or when the water in the wood is in a frozen state. This means that a smaller increase in Θ_m^n during any next step $\Delta\tau$ until reaching of the equality $\Delta\tau \cdot q_{ha} = \Delta\tau \cdot q_{source}$ in these cases is needed. This provides a slower increase in Θ_m^n with an increase of w or when the wood is frozen at the beginning of TTP. Analogously, the larger loading of the autoclave means there is a presence of more heat capacity of the wood in the autoclave and that is why the increase in Θ_m^n in this case is also slower.

4.3 The Common TTP Parametric Model

As it was shown above, the energy consumption of an autoclave cannot be considered as a process with lumped parameters. At the same time, due to the boundary condition (4), it is closely connected with the heating process of timber described by the nonlinear parabolic partial differential Eq. (2), e.g. it is a process with distributed parameters and an infinite dimensional state-space. Using some results from the operator theory implemented for solving PDEs Curtain and Zwart [4], Christofidies [6], Dubljevic and Christofides [13] for model-based control, Eq. (2) and the initial and boundary conditions (3)–(7) can be presented in the following below operation form:

$$\frac{\partial \xi(\omega, t)}{\partial t} = A(\xi(\omega, t), p), \, \forall \omega \in \Omega, t > 0, \, A \in D(A) \tag{14}$$

$$B(\xi(\omega, t), u(t), p) = 0, \forall \omega \in \Omega_B, t > 0 \tag{15}$$

$$\xi(\omega, 0) = \xi_0, \forall \omega \in \Omega \cup \Omega_B \tag{16}$$

$$\theta(\omega, t) = C\xi(\omega, t), \forall \omega \in \Omega \cup \Omega_B, t > 0 \tag{17}$$

Here A and B are nonlinear operators, $\omega = \omega(x, y, z)$ is the spatial coordinate, t is the time, $\xi(\omega, t)$ is the process state variable, p represents the generalized vector of the unknown parameters, upon which our investigation focuses, Ω is the space domain of timber, Ω_B is the space boundary condition, I is the identity, $D(A)$ is domain for defining the operator A [4]. In the infinite-dimensional model described by Eqs. (14)–(17), Eq. (15) is the control boundary condition. For the points located at the geometric boundary, Eqs. (15) and (17) can be rewritten as:

$$B(\xi(\Omega_B), t, \theta_m(t), p) = 0 \tag{18}$$

$$\theta(\Omega_B, t) = \theta_m(t) \tag{19}$$

The vector p of immeasurable initial charging parameters has the following four components:

$$p = p(d, w_0, \theta_0, \gamma), \tag{20}$$

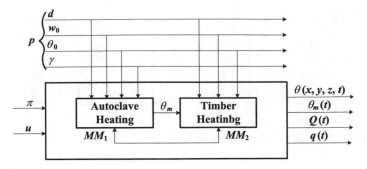

Fig. 4 Scheme of unified TTP model

where d is the equivalent size, w_0, θ_0 are the initial moisture content and the temperature, γ is the relative timber loading charged in the autoclave. All these components are arguments for the dynamic parameters λ_x, λ_y, λ_t, ρ, ρ_B, c in Eq. (2), the initial condition (3) and the boundary conditions (4)–(7).

According to the consideration above in Sect. 4.2, the heat accumulation in the autoclave can be presented in a semi-explicit form:

$$\frac{d\theta_m(t)}{dt} = F_m(\theta_m(t), u(t), p, b), t < T_m, \tag{21}$$

where F_m is a nonlinear function, b is a known vector of physical constants and constructive parameters.

The variations of charging conditions represented by the parametric vector p influence the both interrelated TTP parts as shown in Fig. 4.

Correspondingly to Eqs. (14)–(21) the dynamic behavior of the medium temperature $\theta_m(t)$ must be defined by simultaneous solving this interconnected system of differential and algebraic equations as shown in Fig. 3. The solution in the time interval $0 \leq t \leq T_m$ strongly depends on the parameter-vector p values and it is the only available sensor information in real operations usable for the p reconstruction which we use in this research.

5 Simulation Results

The proposed in this paper approach for the estimation of the unknown parameters is model-based. The core of the research is the developed and practically used first-principle based model of TTP considered above. Using this model, a number of models with artificial neural networks were created as secondary building blocks in the common integrated inference control system.

The solution of the adopted FPM requires pre-assigning the parameter p (Fig. 4). The region of interest (*RoI*) for the formation of the operative situations of autoclave

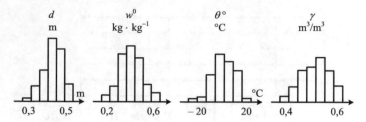

Fig. 5 Histograms of initial charging parameters

charging in our investigation is based on statistical data, gathered expert data the results from special measurements and estimates, obtained following the proposed here approach. Figure 5 shows an example of histograms of all the components of p vector—d, w_0, θ_0, γ.

Due to the large seasonal impact of the initial timber conditions it is reasonable to use relevant seasonal data sets. The accepted for simulation model solutions can be presented in the matrix form:

	u_1^0	u_2^0	u_3^0	u_4^0
π_1	S_{11}	S_{12}	S_{13}	S_{14}
π_2	S_{21}	S_{22}	S_{23}	S_{24}
π_3	S_{31}	S_{32}	S_{33}	S_{34}

$$(22)$$

where π_i are the wood species, u_i^0 is the prescribes heating power, S_{ij} is the set of model solutions for the equation:

$$\theta_{mij}(t) = F_{ij}^m(\pi_i, u_j^0, p_{ij}, b, t), \quad p_{ij} \in \mathbf{P}, \ t \leq T_{m_{ij}}^{\max}; \tag{23}$$

$$q_{ij}(t) = f_{ij}(\pi_i, u_j^0, p_{ij}, b, t); \tag{24}$$

$$Q_{ij}(t) = \int_0^{T_m} q_{ij}(t)dt, \tag{25}$$

where $\theta_m(t)$ is the medium temperature and $q_{ij}(t)$ is the heating flow.

As an illustration of the sets S_{ij} some simulation results are shown in Figs. 6 and 7 for the following conditions: autoclave length 9 m and 2.4 m in diameter, the available heat power $u^0 = 500$ kW, the reference temperature of the heating medium $\theta_m^{\max} = 130\,°C$.

As it was justified above, the impact of some component on the vector parameter p can be very different during the autoclave heating ($t < T_m$) and the timber treatment ($t > T_m$). This is evident from Fig. 8 where the impact of the timber size (d) is opposite to the considered time intervals $(0, T_m)$ and $(0, T)$.

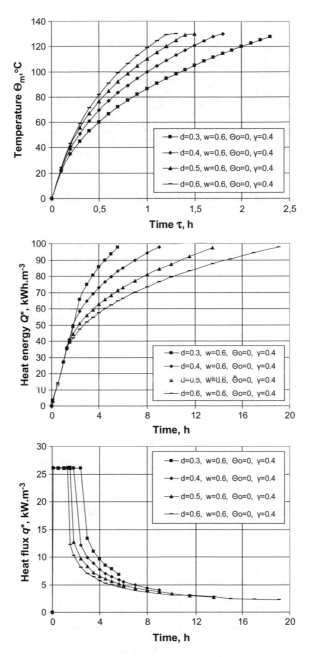

Fig. 6 Change of temperature during T_m and of heat and heat flux during T for non-frozen beech materials with $\Theta_o = 0\ ^\circ C$ and $w = 0.6\ \text{kg.kg}^{-1}$, depending on d at $\gamma = 0.4$

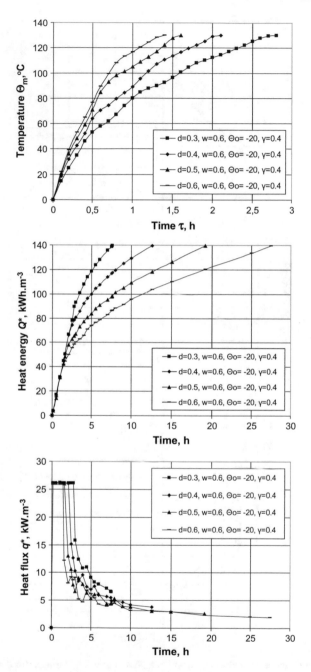

Fig. 7 Change of temperature during T_m and of heat and heat flux during T for frozen beech materials with $\Theta_o = -20\ ^\circ C$ and $w = 0.6\ kg.kg^{-1}$, depending on d at $\gamma = 0.4$

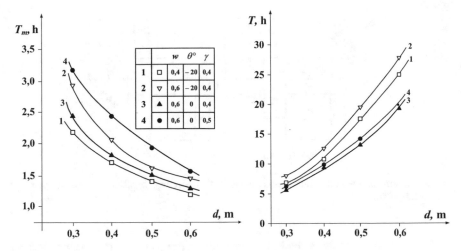

Fig. 8 The impact of timber size on the duration T_m (left) and T (right)

Taking into account the empirical distribution of the vector parameter components (Fig. 5), a sufficient off-line simulation was carried out in order to achieve a reasonable density of the areas of the variables of interest. In cases with considerable changes of the operational conditions the S matrix (22) can be expanded via respective new solutions of the mathematical model in a new *RoI*.

6 Initial Charging Parameter Estimation

6.1 Preliminary Considerations

The estimation of parameter vector $p(d, w_0, \theta_0, \gamma)$ is solved using the sets of parameterized ensembles of explicit according to p solutions of the mathematical model, structured in multidimensional clusters (Fig. 9).

The specific properties of the problem to be solved are:

- All the curves in the clusters S_{ij} are smooth (Figs. 6 and 7);
- Disturbances and noises are small;
- High level conditions κ, ξ, π are predetermined and the acceptable power u_0 is given. This reduces considerably the dimensionality of the search space;
- The initial heating process ($t < T_m$) is relatively fast, but long enough to estimate p in order to determine on time the optimal duration of the j-run T_j^*.;

We use three sources as input data for the estimation procedure:

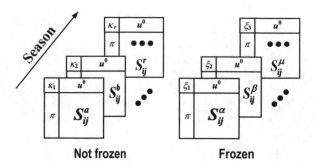

Fig. 9 Clusters of simulated solution for seasons (κ, ξ), operating conditions (π, u_0) and charging parameters $p\ (M_{ij})$

1. The real time profile of the heating medium temperature $\theta_m^e(t)$ and the time T_m to reach the setting temperature θ_m^r. This is the only available sensor data for the TTP.
2. The expert estimates of the operator on duty for the vector p^{op} components $d^{op}, w_0^{op}, \theta_0^{op}, \gamma^{op}$. The values $d^{op}, \theta_0^{op}, \gamma^{op}$ can be accepted as statistically reliable. The personal evaluation of the moisture content w_0 can be with a big deviation from the real value. Thus the operator estimate w_0^{op} is corrected with the statistical data from the previous autoclave runs $\hat{w}_{j-1}$:

$$\hat{w}^{op} = g w_j^{op} + (1 - g)\hat{w}_{j-1}, \tag{26}$$

where the weight g is empirically settled.

The operator estimates are accepted for usage because the initial guess in the search and optimizing procedures are very important and in cases with a lot of local minima—of critical importance.

3. The TTP model-based parameterized solutions (Fig. 9):

$$\theta_m^M(t) = F_m^M\big(\kappa, \xi, \pi, u_0, p^M, t\big), \tag{27}$$

where p^M corresponds to the *RoI*-based cluster of the current j-run.

6.2 Peculiarities of the Proposed Estimation Procedure

- The estimation of the p-vector is based on very scare objective information from three independent sources considered above;
- Only the dynamic data in the time interval $0 < t < T_m$ are used;

- The estimation of the initial charging parameters is organized as a three-staged procedure containing:

  - Simulation to obtain the *RoI* clusters,
  - Search in the relevant cluster space,
  - Optimization using nonlinear programming (NLP);

- *p*-estimation combines both procedures over the parameterized curves like in Eq. (27) and the numerical values of the parameters during the iterations as well;
- A validation procedure is stipulated in order to confirm the consistency of the received *p*-vector estimates on the base of the partial information from the only available first TTP stage with the global-model parameters valid for the whole process;
- The main part of the calculations is performed off-line. The on-line part is with acceptable computer time consumption;
- For the transitions 'parameters-solution' there are used artificial neural networks (NN) after a relevant learning based on the available simulation results.

6.3   Estimation Procedure of P-Vector

The complete scheme of the estimation procedure is presented in Fig. 10. It incorporates some interconnected sub-procedures.

6.3.1   Preliminary Actions

1. Off-line generation of the clusters of the mathematical model solutions (27) spanning all *RoI*.
2. The operator estimates for the initial parameter charging p_0^{op} and particularly the correction of the moisture contents w_0^{op} following Eq. (26).
3. The learning set of dynamic neural networks NN_1 for the given global parameters κ, ξ, π, u_0 addressed to the operational parameters p^M according to *RoI* for the time interval $t < T_m$.
4. The learning set of static neural networks NN_2 under the same conditions for the whole time interval $0 \le t \le T$.

6.3.2   Estimation of *p*-vector on the base of *N*-nearest Neighbourhoods

1. Define the mean squared measure as a distance between the experimental transient function $\theta_m^M(t, p^M)$ (Eq. 27) as shown in Fig. 11.

$$\Delta\theta_m(t) = \theta_m^e(t) - \theta_{m_j}^M(t, p_j^M); \tag{28}$$

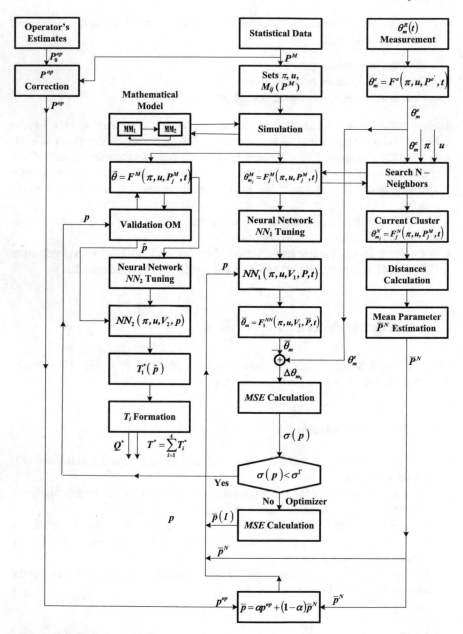

Fig. 10 Scheme of p-vector estimation procedure

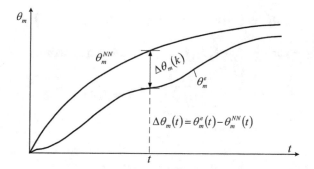

Fig. 11 Evaluation of the distance between experimental $\theta_m^e(t)$ and the arbitrary simulated $\theta_m^M(t, p^M)$ functions

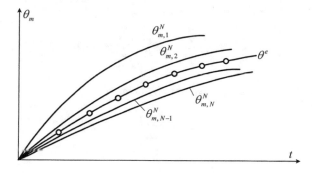

Fig. 12 The family of N nearest neighbours to the experimental transient function $\theta_m^e(t)$

$$D_m^N\left(p_j^M\right) = \frac{1}{T_m} \int_0^{T_m} \left(\theta_m^e(t) - \theta_{m_j}^M\left(t, p_j^M\right)\right)^T E\left(\theta_m^e(t) - \theta_{m_j}^M\left(t, p_j^M\right)\right) dt = \sigma_m^N\left(p_j^M\right)$$

(29)

where E is the diagonal weight matrix.

2. Define the region of interest (RoI) as a threshold σ_m^{Nr} to belong among the nearest N-neighbours.
3. Search the N-nearest neighbours for which the following below inequality is satisfied:

$$\sigma_m^N\left(p_j^M\right) \le \sigma_m^{Nr}, (j = 1, N).$$

(30)

The family of N closest neighbours $\theta_m^N\left(p_j^M, t\right)$ to $\theta_m^e(t)$ is presented in Fig. 12.

4. Derive the mean vector $\bar{p}^N$ for the whole N-family. As p_j^M has the explicit form:

$$p_j^M = \left(p_{j1}^M, p_{j2}^M, p_{j3}^M, p_{j4}^M\right) = \left(d_j, w_{oj}, \theta_{oj}, \gamma_i\right) \tag{31}$$

for each component p_{ji}^M ($j = 1, N, i = 1, 4$) there can be written:

$$\bar{p}_i^N = \frac{1}{N} \sum_{j=1}^N p_{ji}^M, (i = 1, 4) \tag{32}$$

and

$$\bar{p}^N = \left(\bar{p}_1^N, \bar{p}_2^N, \bar{p}_3^N, \bar{p}_4^N\right). \tag{33}$$

5. Derive the weighted mean vector $\bar{\bar{p}}^N$ such that:

$$\bar{\bar{p}}_i^N = \frac{1}{N} \sum_{j=1}^N \beta_j p_{ji}^M, \quad \sum_{j=1}^N \beta_j = 1, \tag{34}$$

where $\beta_j (j = 1, \ldots, N)$ are the weight coefficients:

$$\beta_j = \frac{\sigma_j}{\sum_{j=1}^N \sigma_j} \tag{35}$$

and:

$$\bar{\bar{p}}^N = \left(\bar{\bar{p}}_1^N, \bar{\bar{p}}_2^N, \bar{\bar{p}}_3^N, \bar{\bar{p}}_4^N\right). \tag{36}$$

6. Define the closest neighbour from the N-family possessing the smallest distance $\sigma_m^N\left(p_c^N\right)$ in accordance with Eq. (29) with vector parameter p_c^N.

6.3.3   Formation of a Set of Parameters p to be used as An Input to the Neural Network NN_1

We can use the set of candidate p-vectors as it follows below:

(a) A vector p_c^N of the nearest curve from the N-family;
(b) A mean vector $\bar{p}^N$ from Eq. (33);
(c) A weighted vector $\bar{\bar{p}}^N$ from Eq. (36);
(d) The combined p-vectors using the operator estimates p^{op} and p^N estimates from the N-family:

$$\bar{p} = \alpha p^{op} + (1 - \alpha) p^N. \tag{37}$$

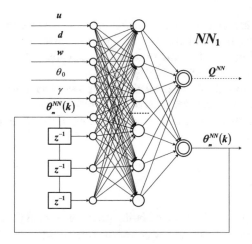

Fig. 13 Dynamic neural network NN_1

Here α is the weight coefficient, defined empirically.

6.3.4 Estimation of *P*-Vector Without Optimization

1. A threshold σ^r is adopted as a requirement for the closeness between the experimental transient profile of the medium temperature $\theta_m^e(t)$ and each candidate-decision of the solution with the received *p*-vectors– p_c^N, $\bar{p}^N$, $\bar{\bar{p}}^N$, $\bar{p}$.
2. If the closest neighbour from the N-family satisfies the condition $\sigma_c^N < \sigma^r$ then the search is successfully finished and the model solution $\theta_m^M\left(\pi, u_0, p_c^M, t\right)$ is directed towards the validation block.
3. In the opposite case $(\sigma_c^N > \sigma^r)$ all received *p*-vectors $(\bar{p}^N, \bar{\bar{p}}^N, \bar{p})$ are used as input data for the dynamic neural network NN_1 (Fig. 13) in order to obtain the corresponding candidate-solutions as a transient curve.
4. For each NN-based solution $\theta_m^M\left(\pi, u_0, p_j, t\right)$ the squared error $\sigma_{m,j}^M$ is calculated using Eq. (29).
5. If some of the candidate-decisions meet the requirement $\sigma_{m,j}^N < \sigma^r$ then the closest of them is fixed as the best and it is directed towards validation.

In the opposite case this best modeled transient solution is accepted as an initial for an iterative optimization procedure

6.3.5 Estimation of *p*-vector with Optimization

1. The minimum square error MSE (Larsen and Marx [42] similar to Eq. (29) is accepted as an objective function but with changeable values of the *p*-vector during the optimization procedure.

2. As the MSE is calculated using experimental $\theta_m^e(t)$ and modeled $\theta_m^{NN}(p, t)$ temperature profiles, in each iteration step e the optimizer sends the current values of the p-vector $p(e)$ to the input of the neural network NN_1 which generates the corresponding transient curve $\theta_m^{NN}(p(e), t)$.
3. Due to the previous stages of the estimation process the initial conditions in the iteration process $p(0)$ can be expected as appropriate.
4. When the requirement $\sigma(p(e)) < \sigma^r$ is accomplished then the vector $p(e)$ is sent for validation.

6.3.6   Validation

1. The validation process is included in the proposed model parameters estimation procedure because a part of the p-vector components (e.g. the cited timber) impacts on the process in a very different way during the stages of the autoclave heating and the timber thermal treatment (Fig. 8).
2. In this work the validation is fulfilled in accordance with the mean mass temperature $\bar{\theta}$ of the timber during the whole TTP process.
3. As sensor data for the state of the timber are unavailable the validation is carried out using the mathematical model-based solution inferring the mean timber temperature.
4. The validation is accomplished on the same family of the model solution used for learning the static neural network NN_2.
5. After the successful validation the accepted final values of the p-vector—$\hat{p}$ are used to define the optimal durations T_i ($i = 1, 4$) of the corresponding stages of the TTP.

7   Intelligent Suboptimal Inference Control of TTP

In conformity with the basic investigations of the distributed-parameter-systems (DPS) time-optimal-control (Lions [1]) the strict optimal control action has a form presented in Fig. 13.

As more than 95% of the time the temperature θ_m must be kept in the maximal available value θ_m^{max} and it is impossible to fulfill the strict re-switching $\theta_m^{max} - \theta_m^{min} - \theta_{nn}^{max}$ due to the cooling/heating inertness, there can be used a slightly suboptimal control with active heating during the time:

$$T_1^* = 1.03 T_{10}^*, \tag{38}$$

where T_{10}^* is the first switching time in a strict time-optimal control Hadjiski and Deliiski [22]. For each vector parameter P there is a special value of π and the power

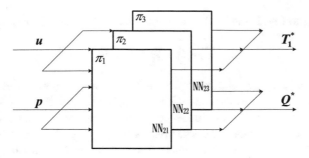

Fig. 14 Static neural network NN$_2$

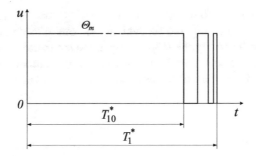

Fig. 15 Minimum time optimal control

u. Yet the value of the heating time T_1^* has been calculated during the initial off-line simulation, accomplishing the optimization procedures:

$$T_1^* = T_1^*(\pi, u, P), \tag{39}$$

using the set $T_1^*(\pi, u, P)$ and a static neural network NN_2 (Fig. 14) has been learned (Fig. 15).

As the specific value of π and the power u^0 are prescribed, the only vector P_j represents the unknown input in the current autoclave j-run necessary to define the optimal heating time $T_1^*(P_j)$.

Following the proposed here system for P_j identification represented in Fig. 10, one of the values $\bar{P}(0)$ or $\bar{P}(l)$ will be available as the input to the neural network NN_2 after finishing the estimation procedure. As the duration T_m of the first TTP stage is many times shorter than the expected whole heating time T, the derived from the neural network NN_2 optimal value $T_1^*(P)$ will be accessible for the control system with a big prediction.

According to Fig. 2, the total operational time T considers four components:

$$T = T_1 + T_2 + T_3 + T_4, \tag{40}$$

where T_1 is the heating time during the supply of steam from the heat generator, T_2 is the time for isochoric heating without supply of steam from the generator, T_3 and T_4 are the times of cooling timber up to an atmospheric pressure and of the air conditioning of the steamed materials respectively. All partial times T_i depend on the different components of the parameter vector P. when P is calculated the current values of $T_i(P)$ can be calculated according to the results, represented in Deliiski and Dzurenda [40].

Figure 16 presents the full scheme of the proposed system for TTP control fulfilled in an autoclave (Fig. 1) taking into account all previous considerations in this work as our previous results. In Fig. 17 BB_i correspond to the building blocks ($i = 1, 4$).

Two main tasks are solved with this system: (i) the estimation of the unknown parameter $P(d, w, \theta_0, \gamma)$ and (ii) the time suboptimal control of the thermal treatment process (TTP).

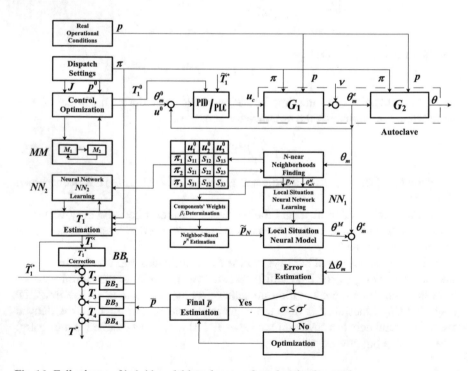

Fig. 16 Full scheme of hybrid model-based system for suboptimal control

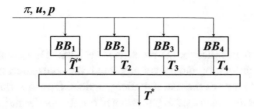

Fig. 17 Scheme of the total operational time T calculation using building blocks BB$_i$

8 Implementation Aspects

More than 20 systems have been implemented for suboptimal TTP time control in different schemes' realizations for autoclaves in Bulgarian and Slovak SME Deliiski and Dzurenda [40]. The heat consumption was reduced 2–3 times. It is possible to achieve a considerable additional economic effect in the new proposed control system with a reconstruction of the unknown initial charging parameters $P(d, w, \theta^0, \gamma)$. As an example of the impact of the model, the boundary, initial and operational uncertainties on the total production time $T = T_1 + T_2 + T_3 + T_4$ (Fig. 15), some results are shown in Table 1 where the situation № 1 is accepted as true.

From Table 1 it is obvious that the incorrectly operator-based estimation of the initial charging conditions P_0^{op} can be a result from large deviations in both sides with unacceptable losses for industrial indexes (capacity, operational costs, quality of timber thermal treatment). The proposed integrated inference intelligent control system possesses a big potential for improvements, according to the results of this research.

Table 1 Change in the total production time T of TTP, depending on d, w, θ_0, and γ

№	P				T		
	d	w	θ_0	γ	T	ΔT	$\Delta T / T$
	m	kg.kg^{-1}	°C	m$^3/$m^3	h	h	%
1	0.4	0.4	0	0.4	12.5	0	0
2	0.3	0.6	0	0.4	9.1	−3.4	−27
3	0.4	0.4	−20	0.4	15.0	2.5	20
4	0.5	0.6	0	0.5	17.2	4.7	38
5	0.4	0.6	−20	0.4	16.1	3.6	29
6	0.3	0.6	0	0.5	9.4	−3.1	−25

9 Conclusions

The effective control of distributed parameter systems (DPS) with shortage or lack of target outputs and/or boundary manipulated variables measurements critically depends on the accuracy of the mathematical models. Due to variations of the operational charging conditions for each particularly run the mathematical models of TTP must be adapted. To reach this goal a hybrid approach is proposed. Some scarce but independent direct sources of information are used—sensor data available only during the short first TTP stage, structured solutions of first principle TTP models obtained for appropriate combination of initial charging conditions, operator's expert estimates and a number of statistical data.

A multivariate approach for successive candidate-vector of estimated model parameters choice is accepted-from the closest transient time profile, through averaging the N-nearest neighbors or nonlinear optimization. A number of neural networks clusters are implemented in order to reduce the approximate models dimension and "parameter-solution" transitions fulfilling without turn to direct simulation. To justify the adopted vector of model parameters a validation module is included in estimation procedure using the simulation results of whole trajectories of the wood materials' average mass temperature. The main part of the computational load could be realized off-line which make the proposed system suitable for industrial application in SMS factories.

References

1. Lions, J.L., Optimal Control of Systems Governed by Partial Differential Equations, Springer-Verlag, 1971.
2. Butkovskii, A.G., Methods of Distributed Parameter Systems Control, Moscow, Nauka, 1975 (in Russian).
3. Ray, W.H., Lainiotis, D.G., Distributed Parameter Pystems, Marcel Dekker, New York, 1978.
4. Curtain, R.F., and Zwart, H.J. (1995). An Introduction to Infinite Dimensional Linear System Theory. Springer.
5. K. Morris, "Control of systems governed by partial differential equations," in *The Control Handbook*. Boca Raton: CRC Press, 2010.
6. P. D. Christofides, Nonlinear and Robust Control of PDE Systems: Methods and Applications to Transport-Reaction Processes, Birkhäuser,Boston, 2001.
7. D. Zheng, K. A. Hoo, System identification and model-based control for distributed parameter systems, Computers & Chemical Engineering,28 (8), (2004) 1361–1375.
8. W. Xie, C. Theodoropoulos, An off-line model reduction-based technique for on-line linear mpc applications for nonlinear large-scale distributed systems, Computer Aided Chemical Engineering 28 (2010), 409–414.
9. P. Dufour, Y. Tour´e, D. Blanc, P. Laurent, On nonlinear distributed parameter model predictive control strategy: on-line calculation time reduction and application to an experimental drying process, Computers & Chemical Engineering 27 (11) (2003) 1533–1542.
10. S. Shvartsman, C. Theodoropoulos, R. Rico-Martinez, I. Kevrekidis, E. Titi, T. Mountziaris, Order reduction for nonlinear dynamic models of distributed reacting systems, Journal of Process Control 10 (2–3) (2000),177–184.

11. E. Balsa-Canto, A. A. Alonso, J. R. Banga, Reduced-order models for nonlinear distributed process systems and their application in dynamic optimization, Industrial & Engineering Chemistry Research 43 (13), (2004) 3353–3363.
12. Brosilow, C., B. Joseph. Technologies of Model Based Control. New Jersey, Prentice Hall, 2002.
13. S. Dubljevic, P. D. Christofides, Predictive control of parabolic PDEs with boundary control actuation, Chemical Engineering Science 61 (18), (2006), 6239–6248.
14. E. Aggelogiannaki, H. Sarimveis, D. Koubogiannis, Model predictive control in long ducts by means of a neural network approximation tool, Applied Thermal Engineering 27 (14–15) (2007) 2363–2369.
15. Z. Hidayat, R. Babuska, B. De Schutter, and A. Nunez, "Observers for linear distributed-parameter systems: A survey," *Proceedings of the 2011 IEEE International Symposium on Robotic and Sensors Environments (ROSE 2011)*, Montreal, Canada, pp. 166–171, Sept. 2011.
16. J. H. Lilly, "Finite-dimensional adaptive observers applied to distributed parameter systems," *IEEE T. Automat. Contr.*, vol. 38, no. 3, pp. 469–474, 1993.
17. R. Curtain,M. Demetriou, and K. Ito, "Adaptive observers for slowly time varying infinite dimensional systems," in *Proc. 37th IEEE Decis. Contr.*, vol. 4, Tampa, FL, USA, Dec. 1998, pp. 4022 –4027.
18. A. Smyshlyaev and M. Krstic, "Backstepping observers for a class of parabolic PDEs," *Syst. & Control Lett.*, vol. 54, no. 7, pp. 613–625, 2005.
19. R. Miranda, I. Chairez, and J. Moreno, "Observer design for a class of parabolic PDE via sliding modes and backstepping," in *Proc. Int. Workshop Var. Struc. Syst.*, Mexico City, Mexico, Jun. 2010, pp. 215–220.
20. M. García, C. Vilas, J. Banga, A. Alonso, Exponential observers for distributed tubular (bio) reactors, AIChE Journal 54 (11) (2008) 2943–2956.
21. M. R. García, C. Vilas, J. R. Banga, A. A. Alonso, Optimal field reconstruction of distributed process systems from partial measurements. Industrial & Engineering Chemistry Research 46 (2) (2007) 530–539.
22. Hadjiski M., N. Deliiski, Advanced Control of the Wood Thermal Treatment Processing, Cybernetics and Information Technologies, Volume 16, No 2, 2016.
23. E. Aggelogiannaki, H. Sarimveis, Nonlinear model predictive control for distributed parameter systems using data driven artificial neural network models, Computers & Chemical Engineering 32 (6) (2008) 1225–1237.
24. Shubin, G. S. Drying and Thermal Treatment of Wood. Lesnaya Promyshlennost, 1990, Moscow (in Russian).
25. Videlov, H. Drying and Thermal Treatment of Wood. University of Forestry, 2003, Sofia (in Bulgarian).
26. Hill, C., Wood Modification—Chemical, Thermal and Other Processes, John Wiley, 2006.
27. Estevetes, L. and H. Pereira, Heat Treatment of Wood, Bioresources, 4(1), 2009, pp. 370–404.
28. Pervan, S. Technology for Treatment of Wood with Water Steam. University in Zagreb, 2009 (in Groatian).
29. Khattabi, A., H.P. Steinhagen, Analysis of Transient Non-linear Heat Conduction in Wood Using Finite-difference Solutions. Holz als Roh- und Werkstoff, 51(4): 272–278, 1993, http://dx.doi.org/10.1007/ BF02629373.
30. Khattabi, A., H.P. Steinhagen, Numerical Solution of Two Dimensional Heating of Logs, Holz als Roh und Werkstoff, 50 (7–8): 308-312, 1992, http://dx.doi.org/10.1007/ BF02615359.
31. Khattabi A., H.P. Steinhagen, Update of "Numerical Solution to Two-dimensional Heating of Logs". Holz als Roh- und Werkstoff, 53(1): 93–94, 1995, http://dx.doi.org/10.1007/ BF02716399.
32. Moreno, R., and Devlieger, F. (1990). Influence of Wood Characteristics and of Heat Conditioning Parameters on Heating Time of Logs. *Ciencia e Investigation Forestal*, № 4.
33. Younsi, R., Kocaefe, D., and Kocaefe Y. (2006). "Three-dimensional simulation of heat and moisture transfer in wood," Appl. Therm. Eng.26(11–12), 1274-1285.

34. Younsi, R., Kocaefe, D., Poncsak, S., and Kocaefe Y. (2007). "Computational modelling of heat and mass transfer during the high-temperature heat treatment of wood,"Appl. Therm. Eng.27, 1424–1431.
35. Deliiski, N., Optimal Control of Wood Steaming in Vener Production, Automatic Measurement and Control in Wood Working Industry, IFAC, Pergamon Press, Bratislava, 1988.
36. Deliiski, N., Modeling and Automatic Control of Heat Energy Consumption Required for Thermal Treatment of Logs. Drvna Industrija, 55 (4): 181–199, 2004.
37. Deliiski, N., Computation of the 2-Dimensional Temperature Distribution and Heat Energy Consumption of Frozen and Non-frozen Logs, Wood Research, 54 (3): 67–78, 2009.
38. Deliiski, N., Transient Heat Conduction in Capillary Porous Bodies. In Ahsan A. (ed) Convection and Conduction Heat Transfer. InTech Publishing House, Rieka: 149–176, 2011, http://dx.doi.org/ https://doi.org/10.5772/21424.
39. Deliiski, N., Modelling of the Energy Needed for Heating of Capillary Porous Bodies in Frozen and Non-frozen States. Lambert Academic Publishing, Scholars' Press, Saarbrücken, Germany, 116 p., 2013, http://www.scholars-press.com//system/covergenerator/build/1060.
40. Deliiski, N., and Dzurenda, L. (2010). Modeling of the Thermal Treatment in the Technologies for Wood Processing. University of Forestry, Sofia (in Bulgarian).
41. Hadjiski, M., N. Deliiski. Cost Oriented Suboptimal Control of the Thermal Treatment of Wood Materials. In: Proc. of 16th IFAC Conference on Technology, Culture, and International Stability, 2015, Sozopol, Bulgaria.
42. Larsen, R. J. and Marx, M. L. "An Introduction to Mathematical Statistics and Its Applications" (2012). Prentice Hall.

Interval Type-2 Fuzzy Logic Dynamic Mutation and Crossover Parameter Adaptation in a Fuzzy Differential Evolution Method

Patricia Ochoa, Oscar Castillo and José Soria

Abstract In this paper we consider the Differential Evolution (DE) algorithm by using fuzzy logic to make dynamic changes in the mutation (F) and crossover (Cr) parameters separately, and this modification of the algorithm we can call it the Fuzzy Differential Evolution algorithm (FDE). A comparison of the FDE algorithm using type-1 fuzzy logic and interval type-2 fuzzy logic is performed for a set of Benchmark functions.

1 Introduction

The use of fuzzy logic in evolutionary computing is becoming a common approach to improve the performance of the algorithms [6–8]. Fuzzy logic or multi-valued logic is based on fuzzy set theory proposed by Zadeh in 1965, which helps us in modeling knowledge, through the use of if-then fuzzy rules. The fuzzy set theory provides a systematic calculus to deal with linguistic information that improves the numerical computation by using linguistic labels stipulated by membership functions [4]. Differential Evolution (DE) is one of the latest evolutionary algorithms that have been proposed. It was created in 1994 by Price and Storn, in an, attempt to solve the problem of the Chebychev polynomial. The following year these two authors proposed the DE for optimization of nonlinear and non-differentiable functions on continuous spaces.

Interval type-2 fuzzy sets (IT2 FSs) and type-1 systems [11, 13] have been gaining popularity rapidly in the last decade. The Mendel-John Representation Theorem [12] for IT2 FSs has played an important role. It states that the footprint of uncertainty

P. Ochoa (✉) · O. Castillo · J. Soria
Tijuana Institute of Technology, Tijuana, Mexico
e-mail: ochoa.martha@hotmail.com

O. Castillo
e-mail: ocastillo@tectijuana.mx

J. Soria
e-mail: jsoria57@gmail.com

© Springer International Publishing AG, part of Springer Nature 2019 81
M Hadjiski and K T Atanassov (eds.), *Intuitionistic Fuzziness and Other Intelligent Theories and Their Applications*, Studies in Computational Intelligence 757,
https://doi.org/10.1007/978-3-319-78931-6_5

(FOU) of an IT2 FS is the union of all its embedded type-1 (T1) FSs, including those that are nonconvex and/or sub-normal. This Representation Theorem implies that all these embedded T1 FSs should be considered in deriving new theoretical results for IT2 FSs [10]. However, it must be noted that almost all applications of fuzzy logic systems use only convex and normal FSs. So, using nonconvex and/or sub-normal embedded T1 FSs in IT2 FSs and systems seems controversial. Some researchers have noticed this problem and proposed to use constrained embedded T1 FSs. Garibaldi [9] pointed out that *"conventional type-2 fuzzy sets also suffer from the problem that they contain embedded sets that do not correspond to meaningful concepts."* [2].

The DE algorithm is a stochastic method of direct search, which has proven effective, efficient and robust in a wide variety of applications such as learning of a neural network, a filter design of IIR and aerodynamical optimization. The DE has a number of important features, which make it attractive for solving global optimization problems, among them are the following: it has the ability to handle non differentiable, nonlinear and multimodal objective functions, usually converges to the optimal solution and uses few control parameters, etc.

The DE belongs to the class of evolutionary algorithms that is based on populations. It uses two evolutionary mechanisms for the generation of descendants: mutation and crossover; finally a replacement mechanism, which is applied between the father vector and son vector determining who survive into the next generation. In the literature we can find that there are papers on Differential Evolution (DE) applications that use this algorithm to solve real world problems. To mention a few, of these works we have the following: A Fuzzy Differential evolution method with dynamic adaptation of parameters for the optimization of fuzzy controllers is presented in [16], this work is an antecedent of this paper including some experiments in which the original Differential Evolution algorithm and the proposed Fuzzy Differential Evolution algorithm.

A fuzzy logic control using a Differential Evolution algorithm aimed at modeling the financial market dynamics is presented in [15], another is the design of an optimized cascade fuzzy controller based on differential evolution: proposed in [19], another is eliciting a transparent fuzzy model using differential evolution presented in [14], the Assessment of human operator functional state using a novel differential evolution optimization based adaptive fuzzy model presented in [20], another work is the proposed Fast evolutionary algorithms [21], another is differential evolution for parameterized procedural woody plant models reconstruction [1], and a hybrid differential evolution algorithm for job shop scheduling problems with expected total tardiness criterion [18].

This paper is organized as follows: Sect. 2 shows the concept of the Differential Evolution algorithm, Sect. 3 presents the Benchmark Functions, Sect. 4 describes the Proposed Method, Sect. 5 shows the fuzzy system, Sect. 6 the experiments and Methodology, and Sect. 7 the Conclusions.

2   Differential Evolution Algorithm

The Differential Evolution (DE) algorithm is a nondeterministic technique based on the evolution of a vector population (individuals) of real values representing the solutions in the search space. The generation of new individuals is carried out by the differential crossing and mutation operators. This is a direct search technique employing a parallel population, which are vectors. Differential Evolution (DE) algorithm is an optimization method belonging to the category of evolutionary computation applied in solving complex optimization problems.

The DE is composed as follows:

2.1   Population Structure

The Differential Evolution algorithm maintains a pair of vector populations, both of which contain Np D-dimensional vectors of real-valued parameters [17].

$$P_{x,\,g} = (x_{i,\,g}), \quad i = 0, 1, \ldots, Np, \quad g = 0, 1, \ldots, g_{max} \tag{1}$$

$$x_{i,\,g} = (x_{j,\,i,\,g}), \quad j = 0, 1, \ldots, D - 1 \tag{2}$$

Where:
P_x = current population.
g_{max} = maximum number of iterations.
i = index population.
j = parameters within the vector.
Once the vectors are initialized, three individuals is selected randomly to produce an intermediate population, $P_{v,g}$, of Np mutant vectors, $v_{i,g}$.

$$P_{v,g} = (v_{i,g}), \quad i = 0, 1, \ldots, Np - 1, \quad g = 0, 1, \ldots, g_{max} \tag{3}$$

$$v_{i,g} = (v_{j,I,g}), \quad j = 0, 1, \ldots, D - 1 \tag{4}$$

Each vector in the current population are recombined with a mutant vector to produce a trial population, P_u, the NP, mutant vector $u_{i,g}$:

$$P_{v,\,g} = (u_{i,\,g}), \quad i = 0, 1, \ldots, Np - 1, \quad g = 0, 1, \ldots, g_{max} \tag{5}$$

$$u_{i,\,g} = (u_{j,\,I,\,g}), \quad j = 0, 1, \ldots, D - 1 \tag{6}$$

2.2 Initialization

Before initializing the population, the upper and lower limits for each parameter must be specified. These 2D values can be collected by two initialized D-dimensional vectors, b_L and b_U, in which subscripts L and U indicate the lower and upper limits respectively. Once the initialization limits have been specified a number generator randomly assigns each parameter in every vector a value within the set range. For example, the initial value $(g=0)$ of the jth parameter of the ith vector is:

$$x_{j, i, 0} = \text{rand}_j(0, 1) \cdot \left(b_{j, U} - b_{j, L}\right) + b_{j, L} \tag{7}$$

2.3 Mutation

In particular, the differential mutation adds uses an equation to combine three different vectors chosen randomly to create a mutant vector.

$$v_{i, g} = \mathbf{x}_{r0, g} + F \cdot (\mathbf{x}_{r1, g} - \mathbf{x}_{r2, g}) \tag{8}$$

2.4 Crossover

To complement the differential mutation search strategy, DE also uses uniform crossover. This is sometimes known as a discrete recombination (dual), in particular, DE crosses each vector with a mutant vector:

$$U_{i, g} = (u_{j, i, g}) = \begin{cases} v_{j,i,g} & \text{if } (rand_j(0,1) \leq Cr \text{ or } j=j_{rand}) \\ x_{j,i,g} & \text{otherwise.} \end{cases} \tag{9}$$

2.5 Selection

If the test vector, $U_{i,g}$ has a value of the objective function equal to or less than its target vector, $\mathbf{X}_{i,g}$. It replaces the target vector in the next generation; otherwise, the target retains its place in the population for at least another generation [9].

$$X_{i, g+1} = \begin{cases} U_{i,g} & \text{if } f(U_{i,g}) \leq f(X_{i,g}) \\ X_{i,g} & \text{otherwise.} \end{cases} \tag{10}$$

The process of mutation, recombination and selection are repeated until the optimum is found, or terminating pre criteria specified is satisfied. DE is a simple, but powerful search engine that simulates natural evolution combined with a mechanism to generate multiple search directions based on the distribution of solutions in the

```
Begin
    G=0
    Create a random initial population x_{i,G} ∀i, i = 1,........,NP
    Evaluate f(x_{i,G}) ∀i, i = 1,.......,NP
    For G=1 to MAX_GEN Do
        For i=1 to NP Do
            Select randomly r_1 ≠ r_2 ≠ r_3:
            J_{rand} = randint(1,D)
            For j=1 to n Do
                If (rand_j[0,1) < CR or j = j_{rand})Then
                    u_{i,j,G+1} = x_{r3,j,G} + F(x_{r1,j,G} - x_{r2,j,G})
            Else
                    u_{i,j,G+1} = x_{i,j,G}
                End If
            End For
            If (f (u_{i,G+1}) ≤ f (x_{i,G} )) Then
                x_{i,G+1} = u_{i,G+1}
            Else
                x_{i,G+1} = x_{i,G}
            End If
            G = G+1
        End For
End
```

Fig. 1 Detailed pseudocode of the Differential Evolution algorithm

current population. Each vector i in the population at generation G, can be called at this moment of reproduction as the target vector will be able to generate one offspring, called trial vector ($\mathbf{u}i,G$).

This trial vector is generated as follows: First of all, a search direction is defined by calculating the difference between a pair of vectors $r1$ and $r2$, called "*differential vectors*", both of them chosen at random from the population. This difference vector is also scaled by using a user defined parameter called "$F \geq 0$". This scaled difference vector is then added to a third vector $r3$, called "*base vector*".

As a result, a new vector is obtained, known as the mutation vector. After that, this mutation vector is recombined with the target vector (also called parent vector) by using discrete recombination (usually binomial crossover) controlled by a crossover parameter $0 \leq CR \leq 1$ whose value determines how similar the trial vector will be with respect to the target vector.

There are several DE variants. However, the most known and used is DE/rand/1/bin, where the base vector is chosen at random, there is only a pair of differential vectors and a binomial crossover is used. The detailed pseudocode of this variant is presented as follows in Fig. 1 [14].

Table 1 Benchmark Functions

Function	Search domain	f min
Sphere (f1)	$[-5.12, 5.12]^n$	0
Griewank (f2)	$[-600, 600]^n$	0
Schwefel (f3)	$[-500, 500]^n$	0
Rastringin (f4)	$[-5.12, 5.12]^n$	0
Ackley (f5)	$[-15, 30]^n$	0
Rosenbrock (f6)	$[-5, 10]^n$	0

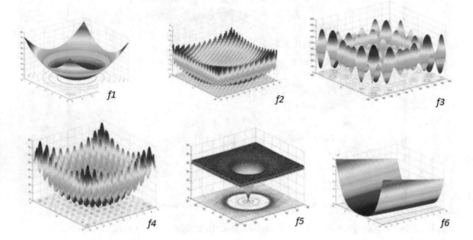

Fig. 2 Benchmark Functions graphs

In the "DE/rand/1/bin" pseudocode rand [0, 1) is a function that returns a real number between 0 and 1. Randint (min, max) is a function that returns an integer number between min and max. *NP*, *MAX GEN*, *CR* and *F* are user-defined parameters and *n* is the dimensionality of the problem [3].

3 Benchmark Functions

In this paper we consider 6 Benchmark functions, which are briefly explained below [5]. Table 1 shows the parameters of the Benchmark functions.

Figure 2 shows the plot of the Benchmark functions.

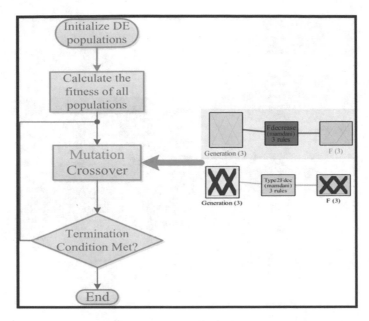

Fig. 3 Differential Evolution with dynamic F parameter

4 Proposed Method

The Differential Evolution (DE) Algorithm is a powerful search technique used for solving optimization problems. In this paper a new algorithm called Fuzzy Differential Evolution (FDE) with dynamic adjustment of parameters is presented.

For this work, four fuzzy systems, where F and Cr are the parameters that are made dynamic, are designed based on type-1 and interval type-2 fuzzy logic. The main goal is to improve the original algorithm dynamically by changing one of the parameters. Figures 3 and 4 show the flowchart of the original algorithm and using type-1 and interval type-2 fuzzy logic.

5 Fuzzy Systems

Four fuzzy systems which are explained separately were developed, two fuzzy systems to modify the F parameter and two fuzzy systems for the Cr parameter.

The type 1 fuzzy system that we are using is defined as follows:

- Contains one input and one output.
- Is Mamdani type.
- All functions are triangular.
- The input of the fuzzy system is the generation.

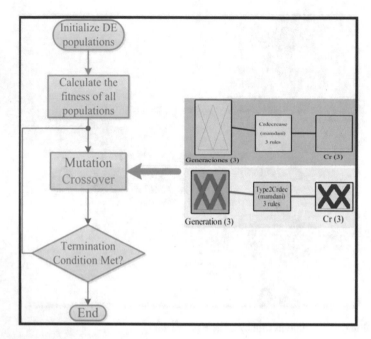

Fig. 4 Differential Evolution with dynamic Cr parameter

1. If (Generation is Low) then (F is High) (1)
2. If (Generation is Medium) then (F is Medium) (1)
3. If (Generation is High) then (F is Low) (1)

Fig. 5 Rules of the fuzzy system for F parameter

- The output of the fuzzy system is the F variable.
- The fuzzy system uses 3 rules and what it does is decrease the F variable in a range of (0.1).

  The rules of the fuzzy system are shown in Fig. 5.
  Figure 6 shows the Type-1 fuzzy system used to dynamically change the F variable.

  The interval type 2 fuzzy system that we are using is defined as follows:

- Contains one input and one output.
- Is Mamdani type.
- All functions are triangular.
- The input of the fuzzy system is generation.
- The output of the fuzzy system is the variable F.
- The fuzzy system uses 3 rules and what it does is decrease the variable F in a range of (0.1).

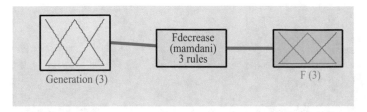

Fig. 6 Type 1 Fuzzy System for F parameter

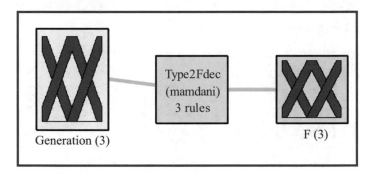

Fig. 7 Interval Type-2 Fuzzy System

1. If (Generation is Low) then (Cr is High) (1)
2. If (Generation is Medium) then (Cr is Medium) (1)
3. If (Generation is High) then (Cr is Low) (1)

Fig. 8 Rules of the fuzzy system for Cr parameter

The rules of the fuzzy system are shown in Fig. 5.

Figure 7 shows the interval type-2 fuzzy system used to dynamically change the F variable.

Figures 8, 9 and 10 show the fuzzy system rules for the Cr parameter and the two type-1 and interval type-2 fuzzy systems, and the fuzzy systems are composed with the same characteristics of the fuzzy systems of the F parameter.

It is noteworthy that the major difference between the fuzzy systems is that in the interval type 2 fuzzy system the membership functions have a degree in the footprint uncertainty.

For this paper the footprint of uncertainty for the interval type 2 fuzzy systems we leave the default value in the triangular function displayed, and this is for all membership functions of the input and output.

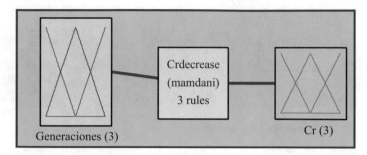

Fig. 9 Type 1 Fuzzy System for Cr parameter

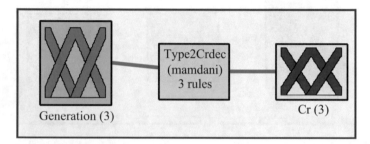

Fig. 10 Interval Type-2 Fuzzy System for Cr parameter

6 Methodology and Experiments

Experiments were made with a set of 6 Benchmark functions of have a global minimum of zero. The Fuzzy Differential Evolution algorithm where the F variable (mutation) and Cr (crossover) change dynamically with the help of fuzzy logic.

Experiments are performed using the first variant of the Fuzzy Differential Evolution algorithm with type 1 fuzzy logic, 30 experiments for each of the functions, which obtain the averages for comparison. Later experiments conducted with the Fuzzy Differential Evolution algorithm using interval type-2 fuzzy logic, where 30 experiments were performed for each function and the average was obtained for each function to make the comparison.

The parameters used in the experiments are defined in Table 2 where D is the number of dimensions, NP is the number of elements in the population, F is the mutation which is dynamic in the type 1 fuzzy system and then the interval type 2 fuzzy system, CR is the variable crossover and G is the number of generations.

Tables 3 and 4 show the comparison results between the Fuzzy Differential Evolution algorithm using type 1 fuzzy logic and Fuzzy Differential Evolution using interval type 2 with F parameter, where the generations the ranging from 100–2000.

Table 3 shows the results in the generation of 100–500 and Table 4 shows the results of generations 1000–2000, for all above Benchmark functions.

Table 2 Parameters the Benchmark functions

Parameter
D = 50
NP = 250
F = Dynamic
CR = 0.1
G = 100–2000
Average of 30 experiments

In Table 4 we can see which of the two shading methods gave better result.

The parameters used in the experiments are defined in Table 5, Cr is the crossover which is dynamic in the type 1 fuzzy system and then the interval type 2 fuzzy system.

Tables 6 and 7 show the comparison results between the Fuzzy Differential Evolution algorithm using type 1 fuzzy logic and Fuzzy Differential Evolution interval type 2 with parameter Cr, where generations ranging from 100–2000.

Table 6 shows the results in the generation of 100–500 and Table 7 shows the results of generations 1000–2000, for all above Benchmark functions.

In Table 7 we can see which of the two methods have better result.

Table 3 Comparison of results for generations from 100 to 500 with parameter F

Comparison between fuzzy logic type 1 and interval type 2

	DE [16]	DE+T1FS	DE+IT2FS	DE [16]	DE+T1FS	DE+IT2FS
G	100	100	100	500	500	500
f1	2.75E+05	1.92E+05	1.95E+01	3.42E+03	1.99E+01	2.17E-03
f2	6.99E+01	4.81E+01	6.91E+01	1.37E+00	5.14E-01	6.21E-01
f3	1.04E+04	1.09E+04	1.11E+04	8.10E+03	4.84E+03	4.83E+03
f4	2.78E+05	1.85E+05	1.46E+03	3.77E+03	3.42E+02	1.64E+02
f5	1.36E+01	1.33E+01	1.33E+01	1.89E+00	2.60E-01	2.66E-01
f6	1.93E+02	1.42E+02	1.44E+02	3.91E+01	3.44E+01	3.62E+01

Table 4 Comparison of results for generations from 1000 to 2000 with parameter F

Comparison between fuzzy logic type 1 and interval type 2

	DE [16]	DE+T1FS	DE+IT2FS	DE [16]	DE+T1FS	DE+IT2FS
G	1000	1000	1000	2000	2000	2000
f1	3.91E+01	2.41E-04	2.53E-08	9.61E-03	3.67E-14	4.23E-18
f2	2.07E-01	1.84E-05	2.54E-05	8.79E-04	1.83E-05	0.00E+00
f3	6388.802	3.96E+00	4.46E+00	4.04E+03	6.36E-04	6.36E-04
f4	2.25E+02	1.51E+02	1.13E+02	6.31E+01	8.47E+01	7.43E+01
f5	2.77E-01	5.98E-04	6.20E-04	1.43E-03	7.53E-09	7.73E-09
f6	1.46E+01	1.55E+00	2.05E+00	5.46E-02	2.20E-10	2.30E-10

Table 5 Parameters the Benchmark functions

Parameter
D = 50
NP = 250
F = 0.1
CR = Dynamic
G = 100–2000
Average of 30 experiments

Table 8 shows the comparison between parameters F and Cr whit type 1 fuzzy logic and interval type 2.

7 Conclusions

We can conclude that the difference between type 1 and intervals type-2 in both cases (F and Cr) is minimal, however the difference that exists between varying the F parameter and the Cr parameter is notorious since we obtain better results by

Table 6 Comparison or results for generations from 100 to 500 with the Cr parameter

Comparison between type 1 fuzzy logic and interval type 2

	DE [16]	DE + T1FS	DE + IT2FS	DE [16]	DE + T1FS	DE + IT2FS
G	100	100	100	500	500	500
f1	2.75E+05	1.26E+03	2.65E+05	3.42E+03	6.48E+01	3.59E+01
f2	6.99E+01	1.32E+00	6.71E+01	1.37E+00	4.65E-02	6.94E-01
f3	1.04E+04	6.57E+03	1.12E+04	8.10E+03	3.91E-02	5.19E+03
f4	2.78E+05	1.85E+03	2.68E+05	3.77E+03	3.42E+02	3.79E+02
f5	1.36E+01	2.78E+00	1.46E+01	1.89E+00	1.36E-01	4.36E-01
f6	1.93E+02	2.20E+01	1.66E+02	3.91E+01	4.69E+00	4.45E+01

Table 7 Comparison or results for generations from 1000 to 2000 with Cr parameter

Comparison between type 1 fuzzy logic and interval type 2

	DE [16]	DE + T1FS	DE + IT2FS	DE [16]	DE + T1FS	DE + IT2FS
G	1000	1000	1000	2000	2000	2000
f1	3.91E+01	2.43E+02	4.82E-04	9.61E-03	6.52E+02	7.29E-14
f2	2.07E-01	2.57E-01	3.42E-05	8.79E-04	4.30E-01	5.22E-15
f3	6.39E+03	6.36E-04	1.45E+01	4.04E+03	6.36E-04	6.36E-04
f4	2.25E+02	1.51E+02	1.44E+02	6.31E+01	8.47E+01	7.64E+01
f5	2.77E-01	3.34E-01	8.44E-04	1.43E-03	6.38E-01	1.05E-08
f6	1.46E+01	3.18E+05	3.02E-01	5.46E-02	3.62E+05	4.32E-09

Table 8 Comparison between F parameter and Cr parameter daynamic with fuzzy logic type 1and interval type 2

Comparison between type 1 fuzzy logic and interval type 2

	DE+T1FS with F	DE+IT1FS with Cr	DE+T2FS with F	DE+IT2FS with Cr
G	2000	2000	2000	2000
f1	3.67E-14	6.52E+02	4.23E-18	7.29E-14
f2	1.83E-05	4.30E-01	0.00E+00	5.22E-15
f3	6.36E-04	6.36E-04	6.36E-04	6.36E-04
f4	8.47E+01	8.47E+01	7.43E+01	7.64E+01
f5	7.53E-09	6.38E-01	7.73E-09	1.05E-08
f6	2.20E-10	3.62E+05	2.30E-10	4.32E-09

dynamically varying the F parameter, Table 8 shows that on average the parameter F is better, since for most functions it was better for both type 1 fuzzy logic and interval type 2.

Although the used fuzzy systems are rather simple, namely they only have one input and one output the results are better than the original Differential Evolution algorithm [16].

We anticipate improving these results with higher dimensions and/or generations. This is because as we know that interval type 2 fuzzy logic has better results with higher uncertainty or noise in the problems. We will also develop more fuzzy systems with more inputs and more outputs to observe the behavior of Fuzzy Differential Evolution algorithm.

Acknowledgements We would like to express our appreciation to CONACYT and Tijuana Institute of Technology for the support provided to this research work.

References

1. A. Zamuda, J. Brest, B. Boˇskoviˊc, V. ˇZumer. "Differential evolution for parameterized procedural woody plant models reconstruction," in ELSEVIER: Applied Soft Computing, Volume 11, Issue 8, December 2011, Pages 4904–4912.
2. D. Wu, A Constrained Representation Theorem for Interval Type-2 Fuzzy Sets Using Convex and Normal Embedded Type-1 Fuzzy Sets, and It's Application to Centroid Computation,presented at the World Conf. Soft Comput., San Francisco, CA", May 2011.
3. E. Mezura-Montes and A. Palomeque-Ortiz, Self-adaptive and Deterministic Parameter Control in Differential Evolution for Constrained Optimization.Efren Mezura-Montes, Laboratorio Nacional de Informatica Avanzada (LANIA A.C.), Rebsamen 80, Centro, Xalapa, Veracruz, 91000, MEXICO 2009.
4. F. Olivas, O. Castillo: Particle Swarm Optimization with Dynamic Parameter Adaptation Using Fuzzy Logic for Benchmark Mathematical Functions. Recent Advances on Hybrid Intelligent Systems 2013: 247–258.

5. F. Valdez, P. Melin, O. Castillo: An improved evolutionary method with fuzzy logic for combining Particle Swarm Optimization and Genetic Algorithms. Applied Soft Computing 11 (2011) 2625–2632.
6. F. Valdez, P. Melin, O. Castillo: Bio-inspired Optimization Methods on Graphic Processing Unit for Minimization of Complex Mathematical Functions. Recent Advances on Hybrid Intelligent Systems 2013: 313–322.
7. F. Valdez., P. Melin., O. Castillo, Evolutionary method combining particle swarm optimization and genetic algorithms using fuzzy logic for decision making, in: Proceedings of the IEEE International Conference on Fuzzy Systems, 2009, pp. 2114–2119.
8. F. Valdez, P. Melin, O. Castillo: Parallel Particle Swarm Optimization with Parameters Adaptation Using Fuzzy Logic. MICAI (2) 2012: 374–385.
9. J. M. Garibaldi, "Alternative forms of non-standard fuzzy sets: A discussion paper," in Type-2 Fuzzy Logic: State of the Art and Future Directions, London, June 2010.
10. J. M. Mendel, "On answering the question "where do i start in order to solve a new problem involving interval type-2 fuzzy sets?"," Information Sciences, vol. 179, no. 19, pp. 3418–3431, 2009.
11. J. M. Mendel, Uncertain Rule-Based Fuzzy Logic Systems: Introduction and New Directions. Upper Saddle River, NJ: Prentice-Hall, 2001.
12. J. M. Mendel and R. I. John, "Type-2 fuzzy sets made simple," IEEE Trans. on Fuzzy Systems, vol. 10, no. 2, pp. 117–127, 2002.
13. L. A. Zadeh, "The concept of a linguistic variable and its application to approximate reasoning-1," Information Sciences, vol. 8, pp. 199–249, 1975.
14. M. Eftekhari, S.D. Katebi, M. Karimi, A.H. Jahanmir: Eliciting transparent fuzzy model using differential evolution, School of Engineering, Shiraz University, Shiraz, Iran, Applied Soft Computing 8 (2008) 466–476.
15. N. Hachicha, B. Jarboui, P. Siarry: A fuzzy logic control using a differential evolution algorithm aimed at modelling the financial market dynamics, Institut Supérieur de Commerce et de Comptabilité de Bizerte, Zarzouna 7021, Bizerte, Tunisia, Information Sciences 181 (2011) 79–91.
16. P. Ochoa, O. Castillo,J. Soria, A Fuzz modelling y Differential evolution method with dynamic adaptation of parameters for the optimization of fuzzy controllers, Norbert Wiener in the 21st Century (21CW), 2014 IEEE Conference on, 24-26 June 2014.
17. Price, R. Storn, Lampinen J. A., Kenneth V, Differential Evolution, Springer 2005.
18. R. Zhang, S. Song, C. Wu. "A hybrid differential evolution algorithm for job shop scheduling problems with expected total tardiness criterion," in ELSEVIER: Applied Soft Computing, Available online 13 March 2012.
19. S.-K Oh. W.-D. Kim, W. Pedrycz, Design of optimized cascade fuzzy controller based on differential evolution: Simulation studies and practical insights, Department of Electrical Engineering, The University of Suwon, Engineering Applications of Artificial Intelligence 25 (2012) 520–532.
20. W. Raofen, J. Zhang, Y. Zhang, X. Wang: Assessment of human operator functional state using a novel differential evolution optimization based adaptive fuzzy model, Lab for Brain-Computer Interfaces and Control, East China University of Science and Technology, Shanghai 200237, PR China, Biomedical Signal Processing and Control 7 (2012) 490–498.
21. X. Yao, Y. Liu, K. H. Liang, Ling G, "Fast evolutionary algorithms. In advances in evolutionary computing: theory and applications," Springer- Verlag, New York, 2003 92v.

Intuitionistic Fuzzy Evaluations for the Analysis of a Student's Knowledge in University e-Learning Courses

Evdokia Sotirova, Anthony Shannon, Taekyun Kim, Maciej Krawczak, Pedro Melo-Pinto and Beloslav Riečan

Abstract In the paper is proposed a method for evaluation of the student's knowledge obtained in the university e-learning courses. For the assessment of the student's solution of the respective assessment units the theory of intuitionistic fuzzy sets is used. The obtained intuitionistic fuzzy estimations reflect the degree of each student's good performances, or poor performances, for each assessment unit. We also consider a degree of uncertainty that represents such cases where the student is currently unable to solve the problem. The method presented here provides the possibility for the algorithmization of the process of forming the student's evaluations.

Keywords e-learning · Intuitionistic fuzzy evaluation · Generalized net modelling

E. Sotirova (✉)
Laboratory of Intelligent Systems, University "Prof. Dr. Assen Zlatarov", Burgas, Bulgaria
e-mail: esotirova@btu.bg

A. Shannon
Resident Fellow, Warrane College, The University of New South Wales, Kensington, NSW, Australia
e-mail: t.shannon@warrane.unsw.edu.au

T. Kim
Division of General Education-Mathematics, Kangwoon University, Seoul, Korea
e-mail: tkkim@kw.ac.kr

M. Krawczak
Systems Research Institute, Polish Academy of Sciences Warsaw School of Information Technology, Warsaw, Poland
e-mail: krawczak@ibspan.waw.pl

P. Melo-Pinto
CITAB, University of Trás-os-Montes and Alto Douro, Vila Real, Portugal
e-mail: pmelo@utad.pt

B. Riečan
Faculty of Natural Sciences, Matej Bel University, Banská Bystrica, Slovakia
e-mail: beloslav.riecan@umb.sk

B. Riečan
Mathematical Institute, Slovak Academy of Sciences, Bratislava, Slovakia

© Springer International Publishing AG, part of Springer Nature 2019
M Hadjiski and K T Atanassov (eds.), *Intuitionistic Fuzziness and Other Intelligent Theories and Their Applications*, Studies in Computational Intelligence 757,
https://doi.org/10.1007/978-3-319-78931-6_6

1   Introduction

Within the context of e-learning, the information exchange between the education and training system and the student is performed electronically. The student obtains information on a given topic at his/her local electronic device. After this the student's acquisition of knowledge can be rated by asking appropriate questions and problems, in order to pass on to the next topic of training.

During the process of e-learning the students have access to different training materials that can be classified as [5]:

- information units to acquire knowledge and skills;
- assessment units—task, problems, test;
- information resources—library, internet, and so on.

The purpose of the present paper is to offer assessments of the process of e-learning within university courses. The research is a continuation of previous investigations of the authors into the modelling of a basic processes and functions of a typical university. In a series of research papers, the authors have studied some of the most important processes of functioning of universities [6–13]. In particular, Generalized Nets, [1, 2], are used to describe the process of student assessment, [6, 10, 11] where the assessments can be represented in an intuitionistic fuzzy form. The concept of Intuitionistic Fuzzy Set was delivered in [3, 4].

The process of evaluation of the problems solved by students is described in [6]. The paper [9] describes the process of evaluation by lecturers of the tasks presented by students. In another paper [10] was constructed a model, which describes the process of evaluation by lecturers. Next, a model that describes the standardization of the process of evaluation by lecturers was constructed [11]. In the next steps of the educational processes there is a modelling of the evaluation of the lecturers themselves [12], and the assessment of the course itself [13]. In [14] the intuitionistic fuzzy assessments are used for modelling the process of e-learning of mathematics topics within university courses.

The aim of the present paper is to use the techniques of intuitionistic fuzzy sets for modelling the process of e-learning within a university educational environment.

2   Proposed Assessment Model

The calculating of the student's knowledge is realized on two phases. First we determine the evaluations of the assessment units for the each student. Then we evaluate the final mark for each student using weight coefficients of the different assessment units and the obtained evaluations for them.

We consider a group of m students, the students are labelled as follows $i = 1, 2, \ldots, m$, and then the students have to be evaluated via j assessment units $j = 1, 2, \ldots, n$. The assessments, which estimate a summative account of the students' knowledge for the different problems, are formed on the basis of a set of intuitionistic fuzzy estimations $\langle \mu, \nu \rangle$ of real numbers from the set $[0, 1] \times [0, 1]$, related to the respective assessment units. These intuitionistic fuzzy estimations reflect the degree of each student's good performances μ, or poor performances ν, for each assessment unit.

The degree of uncertainty $\pi = 1 - \mu - \nu$ represents such cases wherein the student is currently unable to solve the problem and needs additional information. Within the paper the ordered pairs were defined in the sense of intuitionistic fuzzy sets.

2.1 Determination of the Students' Assessments of the Different Units

The way of evaluation of the different units can vary, but for some groups of themes (e.g., mathematics, informatics, physics, chemistry, and etc.), the evaluations of the students' solutions of the different problems can be obtained, in general, by two cases:

Case 1. The assessment unit j contains u^j in number subtasks (questions).

In this case the assessment unit can be in the form of a test with questions with attached possible answers "yes" and "no", or with questions with an attached list of optional answers.

Thus, the evaluation of the i-th student for the j-th assessment unit is obtained by two ways according to the following formula (1), for $i = 1, 2, …, m, j = 1, 2, …, n$.:

$$\langle \mu^j(i), \nu^j(i) \rangle = \left\langle \frac{r^j(i)}{u^j}, \frac{s^j(i)}{u^j} \right\rangle, \tag{1}$$

where:

- $r^j(i)$ is the number of right answers of the subtasks/questions in the assessment unit j,
- $s^j(i)$ is the number of wrong answers of the subtasks/questions in the assessment unit j,
- u^j is the total number of subtasks/questions in the assessment unit j.

Therefore, the degree of uncertainty in this case is determined by the number of the questions which the student had not worked over.

Case 2. The assessment unit j, for example one task, is evaluated independently for w^j levels.

Initially, when there has not been information obtained for the assessment unit, then the estimation is given by the initial values $\langle 0, 0 \rangle$. For $k \geq 0$, the current (k)-st estimation of the i-th student for the j-th assessment unit is obtained on the basis of the previous estimations according to the recurrence relation involved in the following formula (2), $i = 1, 2, …, m, j = 1, 2, …, n$.

$$\left\langle \mu_k^j(i), v_k^j(i) \right\rangle = \left\langle \frac{(k-1) \cdot \mu_{k-1}^j(i) + a_{pl}^j(i)}{k}, \frac{(k-1) \cdot v_{k-1}^j(i) + b_{pl}^j(i)}{k} \right\rangle, \qquad (2)$$

where:

- $\left\langle \mu_{k-1}^j(i), v_{k-1}^j(i) \right\rangle$ is the previous estimation of the j-th assessment unit of the i-th student on the basis of the solutions of the already solved subtasks in the completed levels,
- $\left\langle a_{pl}^j(i), b_{pl}^j(i) \right\rangle$ is the estimation of the level p_l of the j-th assessment unit of the i-th student, for $a_{pl}^j(i), b_{pl}^j(i) \in [0, 1]$, $a_{pl}^j(i) + b_{pl}^j(i) \leq 1$, and $l = 1, 2, \ldots, w$.
- $a_{pl}^j(i)$ and $b_{pl}^j(i)$ are calculated according (3) and (4) in the following way:

$$a_{pl}^j(i) = \begin{cases} \dfrac{c_l^j(i) + d_l^j(i)}{p_l^j}, & \text{if the } i-th \text{ student had worked over level } p_l^j \\ 0, & \text{if the } i-th \text{ student had not worked over level } p_l^j \end{cases}, \qquad (3)$$

$$b_{pl}^j(i) = \begin{cases} \dfrac{p_l^j - (c_l^j(i) + d_l^j(i))}{p_l^j}, & \text{if the } i-th \text{ student had worked over level } p_l^j \\ 0, & \text{if the } i-th \text{ student had not worked over level } p_l^j \end{cases}, \qquad (4)$$

where:

- $c_l^j(i)$ are the points for the solution of the level p_l^j of the j-th assessment unit of the i-th student,
- $d_l^j(i)$ are the points for the description of the decision of the level p_l^j of the j-th assessment unit of the i-th student.

Therefore, the degree of uncertainty, in this case, is equal to 1, when the i-th student did not work over the level p_l^j of the j-th assessment unit.

2.2 Determine of the Final Mark for the i-th Student

Here we introduce intuitionistic fuzzy coefficients $\langle \delta, \varepsilon \rangle$, setting weights of each assessment unit that contribute to the final mark for the i-th student, $i = 1, 2, \ldots, m$. Coefficient δ is based on the number of successive assessment units, and coefficient ε is based on the number of preceding assessment units. An example can clarify this. Suppose, for instance, that a trainee sits for nine assessment units, divided into three levels of difficulty (easy, average, difficult). Let there be four assessment units from the first level, three assessment units from the second level, and two assessment units from the third level. Then the weight coefficients will be distributed as follows:

- from the first level: $< \frac{5}{9}, 0 >$,
- from the second level: $< \frac{2}{9}, \frac{4}{9} >$, and
- from the third level: $< 0, \frac{7}{9} >$.

In this way, the $(j + 1)$st intuitionistic fuzzy estimation $\langle \mu^{j+1}(i), \nu^{j+1}(i) \rangle$, is calculated on the basis of the preceding estimations $< \mu^j(i), \nu^j(i) >$ is obtained according to the following formula (5), for $i = 1, 2,\ldots, m, j = 1, 2,\ldots, n$.

$$\langle \mu^{j+1}(i), \nu^{j+1}(i) \rangle = \left\langle \frac{\mu^j(i) \cdot j + \delta^j \cdot m + \varepsilon^j \cdot n}{j + 1}, \frac{\nu^j(i) \cdot j + \delta^j \cdot n + \varepsilon^j \cdot m}{j + 1} \right\rangle \quad (5)$$

where $<m, n>$ is the estimation of the current assessment unit, $m, n \in [0, 1]$ and $m + n \leq 1$, and $<\delta^j, \varepsilon^j >$ is the weight coefficients of the j-th assessment unit, for $\delta^j, \varepsilon^j \in [0, 1]$, $\delta^j + \varepsilon^j \leq 1$.

The calculated final mark based on all assessment units for the i-th student has to satisfy the necessary "minimal threshold of knowledge". To check this we introduce threshold values: $M_{max}, M_{min}, N_{max}, N_{min}$.

If

$(i) > M_{max}$ & $(i) < N_{min}$,

then the i-th student satisfies the "minimal threshold of knowledge" for the current e-learning course.

If

$(i) < M_{min}$ & $(i) > N_{max}$,

then the i-th student does not satisfy the "minimal threshold of knowledge" for the current e-learning course and he/she has to be evaluated for all assessment units again.

In the rest of the cases the "minimal threshold of knowledge" is undefined and the i-th student has to be evaluated again for the assessment units for which:

$^j(i) \leq M_{max}$ & $\nu \nu^j(i) \geq N_{min \ is}$ valid.

3 Conclusion

In the current paper we have presented the procedure that gives the possibility for the algorithmization of the method of forming the student's evaluations by applying intuitionistic fuzzy estimations. The suggested evaluation methodology and procedures are intended to make the student's evaluations as objective as possible.

In practice, subjective estimation cannot be entirely avoided but it should be made as objective as possible. This can be achieved, to some extent, by approaches which use quantitative methods to utilize the instruments of subjective statistics.

References

1. Atanassov, K., Generalized Nets. World Scientific, 1991.
2. Atanassov, K., On Generalized Nets Theory. Prof. M. Drinov Academic Publishing House, Sofia, 2007.
3. Atanassov, K., Intuitionistic Fuzzy Sets, Springer Physica-Verlag, Berlin, 1999.
4. Atanassov, K., Intuitionistic fuzzy sets. Fuzzy Sets and Systems, Vol. 20, 1986, 87–96.
5. Kensington-Miller, J. Novak, T. Evans, Just do it: flipped lecture, determinants and debate. International Journal of Mathematical Education in Science and Technology, 47 (2016), No.6, 853–862.
6. Melo-Pinto, P., T. Kim, K. Atanassov, E. Sotirova, A. Shannon and M. Krawczak, Generalized net model of e-learning evaluation with intuitionistic fuzzy estimations, Issues in the Representation and Processing of Uncertain and Imprecise Information, Warszawa, 2005, 241–249.
7. Shannon, A., D. Langova-Orozova, E. Sotirova, I. Petrounias, K. Atanassov, M. Krawczak, P. Melo-Pinto, T. Kim, Generalized Net Modelling of University Processes. KvB Visual Concepts Pty Ltd, Monograph No. 7, Sydney, 2005.
8. Shannon, A., K. Atanassov, E. Sotirova, D. Langova-Orozova, M. Krawczak, P. Melo-Pinto, I. Petrounias, T. Kim, Generalized Nets and Information Flow Within a University, Warszawa, 2007.
9. Shannon, A., E. Sotirova, I. Petrounias, K. Atanassov, M. Krawczak, P. Melo-Pinto, T. Kim, Intuitionistic fuzzy estimations of lecturers' evaluation of student work, First International Workshop on Intuitionistic Fuzzy Sets, Generalized Nets and Knowledge Engineering, University of Westminster, London, 6–7 September 2006, 44–47.
10. Shannon, A., E. Sotirova, I. Petrounias, K. Atanassov, M. Krawczak, P. Melo-Pinto, T. Kim, Generalized net model of lecturers' evaluation of student work with intuitionistic fuzzy estimations, Second International Workshop on Intuitionistic Fuzzy Sets, Banska Bystrica, Slovakia, 3 December 2006, Notes on IFS, Vol. 12, 2006, No. 4, 22–28.
11. Shannon, A., E. Sotirova, K. Atanassov, M. Krawczak, P. Melo-Pinto, T. Kim, Generalized Net Model for the Reliability and Standardization of Assessments of Student Problem Solving with Intuitionistic Fuzzy Estimations, Developments in Fuzzy Sets, Generalized Nets and Related Topics. Applications. Vol. 2, System Research Institute, Polish Academy of Science, 2008, 249–256.
12. Shannon, A., D. Dimitrakiev, E. Sotirova, M. Krawczak, T. Kim, Towards a Model of the Digital University: Generalized Net Model of a Lecturer's Evaluation with Intuitionistic Fuzzy Estimations, Cybernetics and Information Technologies, Bulgarian Academy of Sciences, Vol. 9, 2009 No 2, 69–78.
13. Shannon, A., E. Sotirova, M. Hristova, T. Kim, Generalized Net Model of a Student's Course Evaluation with Intuitionistic Fuzzy Estimations in a Digital University, Proceedings of the Jangjeon Mathematical Society, 13 (2010), No 1, 31–38.
14. Sotirova, E., K, Atanassov, A. Shannon, M. Krawczak, Melo-Pinto, B. Riečan, T. Kim, Intuitionistic Fuzzy Evaluations for Analysis of a Student's Knowledge of mathematics in the University e-Learning Courses, 2016 IEEE 8th International Conference on Intelligent Systems, 535–537.

S-Logic with First and Second Imaginary States

Vassil Sgurev

Abstract An imaginary logic, i-logic was introduced for solving of some unsolvable problems in the framework of the classical propositional logic. On the other hand similar unsolvable problems arise in the imaginary logic itself. Introduction of a second imaginary j-logic is suggested in the present work through which this insolvability in the i-logic is surmounted. For this purpose constraints are ushered in between the variables classical r-logic and those of the i-logic and j-logic. This gives rise to complex logics—s, s_1, s_2, respectively, in which functioning of the r, i, and j-logics is being interpreted. It is shown that all these logics are based on the algebraic structures Boolean algebra and lattice. A rule is proposed through which the contradictions may be avoided at the realization of associativity of disjunction and conjunction. A number of results are received for the behavior of conjunction and disjunction in the complex logics being concerned. On their base two truth tables are filled in and shown for the variables of indices r, i, j, s, s_1—for conjunction and disjunction separately. It is pointed out that the logical structures being proposed may be considered as a complex multiple-valued logic with 12 states in which three two valued logics with indices r, i, j are interpreted in an appropriate manner. The field of application of the logical structures being investigated are shown.

Keywords Classical propositional logic · Imaginary logic · Complex s-logic
Conjunction · Disjunction · Lattice · Boolean algebra

1 Introduction

Information technologies, from their rise, are tightly bounded with formal logic. In modern times this connection is getting stronger and stronger [1–3]. The propositional logic lies in the foundations of directions like artificial intelligence, machine

V. Sgurev (✉)
Institute of Information and Communication Tehnologies, Bulgarian Academy of Sciences, Sofia, Bulgaria
e-mail: vsgurev@gmail.com

© Springer International Publishing AG, part of Springer Nature 2019   101
M Hadjiski and K T Atanassov (eds.), *Intuitionistic Fuzziness and Other Intelligent Theories and Their Applications*, Studies in Computational Intelligence 757,
https://doi.org/10.1007/978-3-319-78931-6_7

learning etc. This is especially important for new knowledge inference and intelligent information processing [4–6].

Some logical equations cannot be solved in the framework of this logic. Such is the case when neither of the two possible states—TRUE (T) or FALSE (F) of the set $Q_r = \{T, F\}$ of the logical variable $x \in Q_r$ does not satisfy the relation

$$F \wedge x = T;$$

where $x \in Q_r = \{T, F\}$ [7].

Introduction of imaginary variables $p \in Q_i = \{i, \neg i\}$ is proposed in [3, 8, 9], which may be in one of the states i, and $\neg i$ (not i).

Then the solution may be found by using the state i, namely

$$F \wedge p = F \wedge i = T. \tag{1}$$

2 Real Propositional Logic (*r*-Logic), First Level Imaginary (FL *i*-Logic, or *i*-Logic Only), and the First Level Complex Logic (*s₁*-Logic) Second Level Imaginary Logic (SL *j*-Logic, or *j*-Logic Only) Second Level Complex Logic (*s₂*-Logic)

Here analogically to terminology in the predicate logic the term "first-level imaginary logic" (FL *i*-logic) is introduced, but in this case it concerns *propositional logic*.

The classic propositional logic will be called *real logic* and will be denoted by index r, i.e. *r*-logic.

The imaginary logic will be denoted by index i and will be called *i*-logic. It is defined on the base of the same algebraic structure like the classic propositional logic, namely symmetric idempotent ring, and lattice. Its state i is not equipotent to T, and in the same way $\neg i$ is not equipotent to F. This follows from (1) where the assumption of equipotence of i with T would result in a contradiction.

But when solving logical equations in the framework of the *i*-logic unsolvable cases may also arise like in the *r*-logic, namely $\neg i \wedge p = i; p \in \{i, \neg i\}$.

For solving this case it is expedient a second imaginary logic to be introduced called *j*-logic. It has two contrary states $Q_j = \{j, \neg j\}$. The preceding logical equation may be solved in the following way:

$$\neg i \wedge j = i. \tag{2}$$

Analogically like the *r*-logic and *i*-logic, the *j*-logic is based on the same algebraic structure, namely symmetric idempotent ring, Boolean algebra and lattice [10–12]. Logical variables of the *r*-logic are connected with those of the *i*-logic through (1) and the variables of the *i*-logic are connected with those of the *j*-logic through the novel Eq. (2). No formal connection between the r-logic and j-logic is postulated in advance, i.e. no transitivity between r, i, and j logics is presumed initially.

The behavior in the interaction between the three r, i, and j-logics is investigated in the present work after the introduction of a second imaginary j-logic being proposed.

All results of the classic propositional logic are valid for the real and imaginary logics as they are based on one and the same algebraic structure. Both logics—real and imaginary may be considered as two parallel logics connected by the relation (1) which contains variables from one and the other logics. As both logics are based on one and the same axioms of the lattice then (1) may be considered as additional axiom which interconnects them.

By analogy with the complex numbers $z = a + bi$, where a and b are real numbers and i is the imaginary unit, complex logical variables may be introduced for the relations from (1)

$$g_1 = x_1 \vee p_1; \; g_2 = x_2 \wedge p_2; \tag{3}$$

where $g_1, g_2 \in G = \{g_1, g_1, \ldots g_k, \ldots\}$;

$$x_1, x_2 \in X = \{x_1, x_1, \ldots x_k, \ldots\};$$
$$p_1, p_2 \in I = \{p_1, p_1, \ldots p_k, \ldots\}; \tag{4}$$

where G, X, and I are the sets of variables in the complex, real, and imaginary logic, respectively.

Relations (3) define in essence complex (summary) logic between the real and the imaginary logic which will be further referred to as s-logic.

It is shown in [3, 8, 13, 14] that under the assumptions made, the complex s-logic is up to the axioms of the symmetric idempotent ring, the Boolean algebra, and the lattice, as well as to the additional constraints (1), unifying the real and imaginary logics. All requirements for associativity, idempotence of disjunction and conjunction, as well as distributivity of the disjunction with regard to conjunction, and vice versa, are also observed in the s-logic. At that for each complex variable g the complement $\neg g$ is unique and "symmetry" of the complement exists ($\neg\neg g = g$) and the De Morgan two laws are valid, namely:

$$\neg(g_1 \vee g_2) = \neg g_1 \wedge \neg g_2; \; \neg(g_1 \wedge g_2) = \neg g_1 \vee \neg g_2. \tag{5}$$

It is expedient to define the logical operation "negation" in the s-logic through the relations (5).

All five logical operations of the s-logic—disjunction, conjunction, implication, equivalence, and negation are defined it a way, analogical to the one in i-logic and r-logic. The several following relations, proved in [1, 15] play an important role in the s-logic.

Proposition 2.1 *A relation exists:*
(a)

$$T \wedge i = F \wedge i = T. \tag{6}$$

The following chain of relations is based on Eq. (1)

$$T \wedge i = (F \wedge i) \wedge i = F \wedge (i \wedge i) = F \wedge i = F. \tag{7}$$

The negation of both sides of (6) keeping in mind (7) results in

$$\neg(T \wedge i) = F \vee \neg i = T \vee \neg i = \neg T. \tag{8}$$

Proposition 2.2 *A relation exists:*

$$T \vee i = F \vee i. \tag{9}$$

If the state i in the left hand side the state i is substituted by $i \vee \neg i$ and (8) is taken in mind, then the following chain of equivalences is reached:

$$T \vee i = T \vee (i \vee \neg i) = (T \vee \neg i) \vee i = F \vee i. \tag{10}$$

The negation of both sides of (9) results in:

$$F \wedge \neg i = T \wedge \neg i. \tag{11}$$

It is further important to define in how many and what states the logical variables (3) of the s-logic, real, and imaginary logic may be.

(a) Eight elementary states may be pointed out distributed in the following two groups corresponding to conjunction and disjunction, namely:

$$Q_k = \{(T \wedge i), (F \wedge i), (T \wedge \neg i), (F \wedge \neg i)\};$$
$$Q_d = \{(T \vee i), (F \vee i), (T \vee \neg i), (F \vee \neg i)\},$$

where $Q_s = Q_k \cup Q_d$ is the total number of states of the logical variables (3).

It follows from (6) and (8) that the first two states are equivalent between them and according (1) both are equivalent to T. Last two states of Q_k are equivalent between them and, according (7), they may be represented by $(F \wedge \neg i)$. Analogically it follows from (8) that the last two states in Q_d are equivalent between them and equivalent to T a well and the first two states in the same set are equivalent between them and may be represented by $(T \vee i)$. Hence $Q_k = \{F, (F \wedge \neg i)\}$; $Q_d = \{T, (T \vee i)\}$; $Q_s = \{T, (T \vee i), F, (F \wedge \neg i)\}$.

Table 1 Truth table of real logic—conjunction

x_1	x_2	$x = x_1 \wedge x_2$
T	T	T
T	F	F
F	T	F
F	F	F

Table 2 Truth table of imaginary logic—conjunction

p_1	p_2	$p = p_1 \wedge p_2$
i	i	i
i	$\neg i$	$\neg i$
$\neg i$	i	$\neg i$
$\neg i$	$\neg i$	$\neg i$

In this way only four states remain from the initial eight in the s-logic after the corresponding transformations. Two of them are from the r-logic and two are characteristic for the s-logic only.

(b)   The states which are characteristic of the imaginary logic are only two, namely $Q_i = \{i, \neg i)\}$.

(c)   There are two only states in the real r-logic in the same manner $Q_r = \{T, F\}$.

Then only six states will be contained in general in the s logic, namely $Q = Q_s \cup Q_r \cup Q_i$;

$$Q = \{(T \vee i), (F \wedge \neg i), T, F, i, \neg i\}.$$

It follows from that the s-logic may be considered as a six valued logic which integrates in an entity both logics—real and imaginary through the additional axiom (1)

In the algebraic structure of the S-logic the state $(T \vee i)$ plays the role of the unity "1" and the state $(F \wedge \neg i)$—the role of zero "0". This also follows from the relations:

$$(T \vee i) \vee (F \wedge \neg i) = ((T \vee i) \vee F) \wedge ((T \vee i) \vee \neg i) = (T \vee i) \wedge (T \vee i) = T \vee i; \quad (12)$$

$$(T \vee i) \wedge (F \wedge \neg i) = ((F \wedge \neg i) \wedge T) \vee ((F \wedge \neg i) \wedge i) = (F \wedge \neg i) \vee (F \wedge \neg i) = F \wedge \neg i. \quad (13)$$

2.1   Conjunction Between States of r, FL −i, and s₁ Logics

The two Tables 1 and 2 present the truth tables well known from the propositional logic of the implication in the real and imaginary logic.

Verification of the truth of the second and third rows of Table 3 is done in (13) and the truth of the first and bottom rows of Table 3 follows from the idempotence of conjunction.

Table 3 Truth table for the g-logic—conjunction

g_1	g_2	$g=g_1 \wedge g_2$
$T \vee i$	$T \vee i$	$T \vee i$
$T \vee i$	$F \wedge \neg i$	$F \wedge \neg i$
$F \wedge \neg i$	$T \vee i$	$F \wedge \neg i$
$F \vee \neg i$	$F \wedge \neg i$	$F \wedge \neg i$

Table 4 Truth table for imaginary and real logic—conjunction

p	x	$g=p \wedge x$
i	T	T
i	F	T
$\neg i$	T	$F \wedge \neg i$
$\neg i$	F	$F \wedge \neg i$

Table 5 Truth table of complex and real logic—conjunction

g	x	$g=g \wedge x$
$T \vee i$	T	T
$T \vee i$	F	T
$F \wedge \neg i$	T	$F \wedge \neg i$
$F \wedge \neg i$	F	$F \wedge \neg i$

Table 6 Truth table of imaginary and complex logic—conjunction

p	g	$g=p \wedge g$
i	$T \vee i$	$T \vee i$
i	$F \wedge \neg i$	$F \wedge \neg i$
$\neg i$	$T \vee i$	$F \wedge \neg i$
$\neg i$	$F \wedge \neg i$	$F \wedge \neg i$

The truth of Tables 4, 5 and 6 may be checked taking in mind constraints (6)–(11) by rows respectively:

(a) Table 4

- first row: $(i \rightarrow T) = \neg i \vee T = F$;
- second row: $(i \rightarrow F) = (\neg i \vee F) = (\neg i \vee T) = F$;
- third row: $(\neg i \rightarrow T) = (\neg \neg i \vee T) = (i \vee T)$;
- fourth row: $(\neg i \rightarrow F) = (\neg \neg i \vee F) = (i \vee F) = (T \vee i)$.

(b) Table 5

- first row: $((T \vee i) \rightarrow T) = ((F \wedge \neg i) \vee T) = (T \wedge (T \vee \neg i)) = (T \wedge F) = F$;
- second row: $((T \vee i) \rightarrow F) = ((F \wedge \neg i) \vee F) = (F \wedge (F \vee \neg i)) = (F \wedge (F \vee \neg i)) = F \wedge F = F$;
- third row: $((F \wedge \neg i) \rightarrow T) = ((T \vee i) \vee T) = (T \vee i)$;
- fourth row: $((F \wedge \neg i) \rightarrow F) = ((T \vee i) \vee F) = (T \vee i)$.

(c) Table 6

Table 7 Truth table of real logic—disjunction	x_1	x_2	$x = x_1 \vee x_2$
	T	T	T
	T	F	T
	F	T	T
	F	F	F

Table 8 Truth table of imaginary logic—disjunction	p_1	p_2	$p = p_1 \vee p_2$
	i	i	i
	i	$\neg i$	i
	$\neg i$	i	i
	$\neg i$	$\neg i$	$\neg i$

- first row: $(i \rightarrow (T \vee i)) = (\neg i \vee (T \vee i)) = (T \vee i)$;
- second row: $(i \rightarrow (F \wedge \neg i)) = (\neg i \vee (F \wedge \neg i)) = ((\neg i \vee F) \wedge \neg i) = (F \wedge \neg i)$;
- third row: $(\neg i \rightarrow (T \vee i)) = (\neg \neg i \vee (T \vee i) = (T \vee i)$;
- fourth row: $(\neg i \rightarrow (F \wedge \neg i)) = (\neg \neg i \vee (F \wedge \neg i)) = ((F \vee i) \wedge i) = ((F \wedge i) \vee i) = (T \vee i)$.

2.2 Disjunction Between States of r, FL −1, and s₂ Logics

The results of the disjunction between the real and the imaginary logic are shown in Tables 7 and 8 respectively. The former one includes only the states $\{T, F\}$ and the latter—$\{i, \neg i\}$. The results of the disjunction in the frame of the s-logic only are shown in Table 9 when using its states $\{(T \vee i), (F \wedge \neg i)\}$. The first and last rows of this table immediately follow from the idempotence of the disjunction. The second and third rows lead to the same results ensuing from the commutativity of disjunction.

$$(T \vee i) \vee (F \wedge \neg i) = ((T \vee i) \vee F) \wedge ((T \vee i) \vee \neg i) = (T \vee i) \wedge (T \vee i) = (T \vee i). \quad (14)$$

The results of the disjunction for the case when states of one only logic are used, and are shown in Tables 7, 8 and 9. It is of not less importance results to be achieved for the disjunction when states of the three different logics—r, i, and s logics are used. The results of the disjunction between the different states of the real and imaginary logic are described in Table 10. The results in the first and second rows of this table follow from relations (9) and in the third and fourth rows—from (11).

The disjunction between the states of the S-logic and the real one is reflected in the truth Table 11. The following three chains of equivalences confirm their truth. Relations from (6)–(11) are considered in them.

$$(T \vee i) \vee T = T \vee i; (T \vee i) \vee F = T \vee i;$$

Table 9 Truth table of real and g-logic—disjunction

g_1	g_2	$g=g_1 \vee g_2$
$T \vee i$	$T \vee i$	$T \vee i$
$T \vee i$	$F \wedge \neg i$	$T \vee i$
$F \wedge \neg i$	$T \vee i$	$T \vee i$
$F \wedge \neg i$	$F \wedge \neg i$	$F \wedge \neg i$

Table 10 Truth table of imaginary and real logic—disjunction

p	x	$g=p \vee x$
i	T	$T \vee i$
i	F	$T \vee i$
$\neg i$	T	F
$\neg i$	F	F

Table 11 Truth table of complex and real logic—disjunction

g_1	x	$g=g_1 \vee x$
$T \vee i$	T	$T \vee i$
$T \vee i$	F	$T \vee i$
$F \wedge \neg i$	T	F
$F \wedge \neg i$	F	F

Table 12 Truth table of imaginary and complex logic—disjunction

p	g_1	$g=p \vee g_1$
i	$T \vee i$	$T \vee i$
i	$F \wedge \neg i$	$T \vee i$
$\neg i$	$T \vee i$	$T \vee i$
$\neg i$	$F \wedge \neg i$	$F \wedge \neg i$

$$(F \wedge \neg i) \vee T = T \wedge (T \vee \neg i) = T \wedge F = F;$$
$$(F \wedge \neg i) \vee F = F \wedge (F \vee \neg i) = F \wedge F = F.$$

The results from the disjunction between the states of the imaginary and the S-logic are demonstrated in Table 12. The following three logical chains of equivalence are deduced on the base of relations (6)–(11) that confirm the results of the truth Table 12.

$$i \vee (T \vee i) = T \vee i; \quad \neg i \vee (F \wedge \neg i) = (F \vee \neg i) \wedge \neg i = F \vee \neg i;$$
$$i \vee (F \wedge \neg i) = (F \vee i) \wedge i = (F \wedge i) \vee i = T \vee i;$$
$$\neg i \vee (T \vee i) = T \vee i.$$

The following general conclusions may be drawn when comparing six Tables 7, 8, 9, 10, 11 and 12:

(a) The comparison of the three truth Tables 7, 8 and 9 shows that for the disjunction in them, the requirement of the classic propositional logic is observed—third column contains one "False" and three "True" states.

(b) The results shown in Table 9 for the disjunction between the states $\{(T \vee i),$ $(F \wedge \neg i)\}$ of the S-logic coincide with the results shown in Table 12 of the disjunction of that logic and the imaginary logic. The requirements of the classic propositional logic are also observed in them—one "False" and three "True" states. This is due to the lack of specific constraints similar to those from (11).

(c) The relation in the third row of the truth Table 10 corresponds to the constraints available in (8). It imposes specific constraints between the real and the imaginary logics due to which this result seems unnatural for the disjunction. Two states of two different logics—T and $(T \vee i)$ are received in third column of Table 10. The two other states—$\{T, (F \wedge \neg i)\}$ are not used in this case of the disjunction. At that the lower "bound of False" is from the real logic and the upper "bound of True" is from the s-logic. "Raise" of the "Truth bound" is observed. States of the third, s-logic are encountered in the results of the disjunction between the real and the S-logic due to the constraint (11).

(d) The results of the disjunction between the real and S-logic, reflected in Table 11 are the same as in the previous Table 10—between the real and imaginary logics. The same considerations and results are valid like in item (c).

(e) Comparison of Tables 10, 11 and 12 leads to the conclusion that the results of the disjunction between the real logic and the other two logics are related to some "distortion" of results, described in item (c), due to the constraints (1) and (11). Such effect is not encountered at the disjunction between all states of the imaginary logic and the S-logic which remains in the frame of the classic propositional logic.

It seems expedient to check whether the relations of Tables 1, 2, 3, 4, 5 and 6 at conjunction correspond to Tables 7, 8, 9, 10, 11 and 12 at disjunction by applying the De Morgan laws. Such relations exist for the classic propositional logic. This is of great importance when logical operations are between states of different logics.

This is evident by assumption for Tables 1, 2, 7 and 8 respectively. For each of the remaining Tables 3, 4, 5 and 6 a correspondence will be sought to Tables 9, 10, 11 and 12 by checking a control row of each table:

- Negation of the relations of the first row of Table 3 through the De Morgan laws results in

$$\neg(T \vee i) = \neg(T \vee i) \vee \neg(T \vee i) = (F \wedge \neg i) = (F \wedge \neg i) \vee (F \wedge \neg i) = F \wedge \neg i;$$

which corresponds to the relations of the fourth row of Table 9;

- The same operation for the second row of Table 4 leads to

$$\neg T = \neg(i \wedge T) = \neg i \vee F = T \vee \neg i = F;$$

which corresponds to the third row of Table 10;

- The same operation for the third row of Table 5 results in

$$\neg(F \wedge \neg i) = \neg(F \wedge \neg i) \vee \neg T = (T \vee i) \vee F = T \vee i;$$

  which corresponds to the second row of Table 11;
- The same operation for the fourth row of Table 6 results in

$$\neg(F \wedge \neg i) = \neg\neg i \vee \neg(F \wedge \neg i) = i \vee (T \vee i) = T \vee i;$$

  which corresponds to the first row of Table 12.

Analogic results may be respectively obtained for the other rows of the tables under consideration.

Summing up:

Results achieved for the conjunction and disjunction in the S-logic offer a chance some general conclusions to be made:

(1) Tables 1, 2, 3, 4, 5 and 6 thoroughly define the operation "conjunction" between any pair of the six possible states of the set Q of the S-logic. At that the constraint (1) is considered in Tables 4 and 5 between the real and the imaginary logics.
(2) Disjunction between the six states of the set Q of the S-logic is thoroughly defined in Tables 7, 8, 9, 10, 11 and 12. Constraint (9) between the real and imaginary logic is reflected in Tables 10 and 11 respectively.
(3) Negation $\neg$ in the S-logic is defined by the relations in (5).
(4) Then through the three operations thus defined—conjunction, disjunction, and negation all other relations in the S-logic may be defined—implication and equivalence as well as the other formulae, analogic to those of the classic propositional logic and at that keeping in mind constraint (1), equivalent to the negation (8).

A conclusion may be made from the equipollent relations received in the present work

$$F \wedge i = T \vee F \text{ and } T \vee \neg i = T \wedge F \tag{15}$$

that i is not equivalent to T neither is $\neg i$ equivalent to F. These relations demonstrate that the state i may be considered as "truer" than T and $\neg i$—as "falser" than F. In this sense we may accept that the imaginary logic is "senior" or "more meaningful" than the real one. And although that the real and the imaginary logics are constructed on the same logical structure, their states i and T which correspond to one "1" in the Boolean algebra and $\neg i$ and F corresponding to zero "0" in the same algebraic structure, are different. They are united in the complex S-logic through the relations (1) and (9) and its two states $(T \vee i)$ corresponding to one and $(F \wedge \neg i)$ corresponding to zero.

3 Specific Features of the Interaction Between r-, i-, j-, s_1-, and s_2-Logics

In Sect. 2 the two imaginary logics—i-logic and j-logic, were introduced, as well as the two complex logics—s_1-logic and s_2-logic. In this section we will consider the interrelations between the two imaginary, the two complex logics.

$$q_1 = p_1 \vee t_1; \quad q_2 = p_2 \wedge t_2; \tag{16}$$

where

$$q_1, q_2 \in Q = \{q_{i1}, q_{i2}, \ldots\}; t_1, t_2 \in G = \{t_{i1}, t_{i2}, \ldots\}; p_1, p_2 \in I = \{p_1, p_2, \ldots\}, \tag{17}$$

where G, Q, and I are the sets of variables in the complex, first, and second imaginary logics respectively.

Relations (16) define in essence complex (summary) logic between the two imaginary logics which will be further referred to as s-logic.

It is shown in [8, 3] that under the assumptions made, the complex s_1-logic is up to the axioms of the symmetric idempotent ring, the Boolean algebra, and the lattice, as well as to the additional constraints (1), unifying the two imaginary logics. All requirements for associativity, idempotence of disjunction and conjunction, as well as distributivity of the disjunction with regard to conjunction, and vice versa, are also observed in the s_1-logic. At that for each complex variable g the complement $\neg q$ is unique and "symmetry" of the complement exists ($\neg\neg q = q$) and the De Morgan two laws are valid, namely:

$$\neg(q_1 \vee q_2) = \neg q_1 \wedge \neg q_2; \neg(q_1 \wedge q_2) = \neg q_1 \vee \neg q_2. \tag{18}$$

It is expedient to define the logical operation "negation" in the s-logic through the relations (18).

All five logical operations of the s_1-logic—disjunction, conjunction, implication, equivalence, and negation are defined it a way, analogical to the one in i-logic. The several following relations, proved in [1, 15] play an important role in the s_1-logic.

Proposition 3.1 *A relation exists:*
(a)

$$i \wedge j = \neg i \wedge j = i. \tag{19}$$

The following chain of relations is based on Eq. (1)

$$i \wedge j = (\neg i \wedge j) \wedge j = \neg i \wedge (i \wedge j) = \neg i \wedge j = \neg i. \tag{20}$$

The negation of both sides of (19) keeping in mind (20) results in

$$\neg(i \wedge i) = \neg i \vee \neg j = i \vee \neg j = \neg i. \tag{21}$$

Proposition 3.2 *A relation exists:*

$$i \vee j = \neg i \vee j. \tag{22}$$

If the state i in the left hand side i is substituted by $j \vee \neg i$ and (21) is taken in mind, then the following chain of equivalences is reached:

$$i \vee j = i \vee (j \vee \neg j) = (i \vee \neg j) \vee j = \neg i \vee j. \tag{23}$$

The negation of both sides of (9) results in:

$$\neg i \wedge \neg j = i \wedge \neg j. \tag{24}$$

It is further important to define in how many and what states the logical variables (16) of the s_1-logic, real, and imaginary logic may be.

(a) Eight elementary states may be pointed out distributed in the following two groups corresponding to conjunction and disjunction, namely:

$$Q_k = \{(i \wedge j), (\neg i \wedge j), (i \wedge \neg j), (\neg i \wedge \neg j)\};$$
$$Q_d = \{(i \vee j), (\neg j \vee j), (i \vee \neg j), (\neg i \vee \neg j)\},$$

where $Q_s = Q_k \cup Q_d$ is the total number of states of the logical variables (16).

It follows from (19) and (21) that the first two states are equivalent between them and according (1) both are equivalent to i. Last two states of Q_k are equivalent between them and, according (7), they may be represented by $(F \wedge \neg i)$. Analogically it follows from (21) that the last two states in Q_d are equivalent between them and equivalent to i a well and the first two states in the same set are equivalent between them and may be represented by $(i \vee j)$. Hence

$$Q_k = \{\neg i, (\neg i \wedge \neg j)\}; \quad Q_d = \{i, (i \vee j)\}; \quad Q_s = \{i, (i \vee j), \neg i, (\neg i \wedge \neg j)\}.$$

In this way only four states remain from the initial eight in the s-logic after the corresponding transformations. Two of them are from the *r*-logic and two are characteristic for the *s*-logic only.

(b) The states which are characteristic of the imaginary logic are four, namely
   $Q_j = \{j, \neg j\}$.
(c) There are two only states in the s_1-logic in the same manner $Q_r = \{i, \neg i\}$.

Then only six states will be contained in general in the s_1-logic, namely $Q = Q_{s1} \cup Q_i \cup Q_j$;

$$Q = \{(T \vee i), (F \wedge \neg i), T, F, i, \neg i\}.$$

Table 13 First order imaginary logic—conjunction	p_1	p_2	$p = p_1 \wedge p_2$
	i	i	i
	i	$\neg i$	$\neg i$
	$\neg i$	i	$\neg i$
	$\neg i$	$\neg i$	$\neg i$

Table 14 Second order imaginary logic—conjunction	t_1	t_2	$t = t_1 \wedge t_2$
	j	j	j
	j	$\neg j$	$\neg j$
	$\neg j$	j	$\neg j$
	$\neg j$	$\neg j$	$\neg j$

Table 15 First and second order imaginary logic—conjunction	q_1	q_2	$q = q_1 \wedge q_2$
	$i \vee j$	$i \vee j$	$i \vee j$
	$i \vee j$	$\neg i \wedge \neg j$	$\neg i \wedge \neg j$
	$\neg i \wedge \neg j$	$i \vee j$	$\neg i \wedge \neg j$
	$\neg i \vee \neg j$	$\neg i \wedge \neg j$	$\neg i \wedge \neg j$

It follows from that the s_1-logic may be considered as a six valued logic which integrates in an entity both logics.

In the algebraic structure of the s_1-logic the state $(i \vee \neg j)$ plays the role of the unity "1" and the state $(\neg i \wedge \neg j)$—the role of zero "0". This also follows from the relations:

$$(i \vee j) \vee (\neg i \wedge \neg j) = ((i \vee j) \vee \neg i) \wedge ((i \vee j) \vee \neg j) = (i \vee j) \wedge (i \vee j) = i \vee j; \quad (25)$$

$$(i \vee j) \wedge (\neg i \wedge \neg j) = ((\neg i \wedge \neg j) \wedge i) \vee ((\neg i \wedge \neg j) \wedge j) = (\neg i \wedge \neg j) \vee (\neg i \wedge \neg j) = \neg i \wedge \neg j. \quad (26)$$

3.1 Conjunction

The two Tables 13 and 14 present the truth tables well known from the propositional logic of the two imaginary logics.

Verification of the truth of second and third rows of Table 15 is done in (26) and from the idempotence the truth of the first and bottom rows of Table 15 follows.

Table 16 First order imaginary logic—disjunction	p_1	p_2	$p=p_1 \vee p_2$
	i	i	i
	i	$\neg i$	i
	$\neg i$	i	i
	$\neg i$	$\neg i$	$\neg i$

Table 17 Second order imaginary logic—disjunction	t_1	t_2	$t=t_1 \vee t_2$
	j	j	j
	j	$\neg j$	j
	$\neg j$	j	j
	$\neg j$	$\neg j$	$\neg j$

Table 18 First and second order imaginary logics	q_1	q_2	$q=q_1 \vee q_2$
	$i \vee j$	$i \vee j$	$i \vee j$
	$i \vee j$	$\neg i \wedge \neg j$	$i \vee j$
	$\neg i \wedge \neg j$	$i \vee j$	$i \vee j$
	$\neg i \wedge \neg j$	$\neg i \wedge \neg j$	$\neg i \wedge \neg j$

3.2 Disjunction

The results of the disjunction between the imaginary logics are shown in Tables 16 and 17 respectively. The former one includes only the states $\{i, \neg i\}$ and the latter—$\{j, \neg j\}$. The results of the disjunction in the frame of the s_1-logic only are shown in Table 18 when using its states $\{(i \vee j), (\neg i \wedge \neg j)\}$. The first and last rows of this table immediately follow from the idempotence of the disjunction. The second and third rows lead to the same results ensuing from (25).

$$(i \vee \neg i) \vee (j \wedge \neg j) = ((T \vee i) \vee F) \wedge ((T \vee i) \vee \neg i) = (T \vee i) \wedge (T \vee i) = (T \vee i).$$

The results of the disjunction for the case when states of one only logic are used, and are shown in Tables 16, 17 and 18. It is of not less importance results to be achieved for the disjunction when states of the three different logics—i, j, and s_1 logics are used.

The verification of the second and third rows of Table 18 is done in (25) and the truth of the first and bottom rows of the same table follows from the idempotence.

The comparison of the three truth Tables 16, 17 and 18 shows that for the disjunction in them, the requirement of the classic propositional logic is observed – third column contains one "False" and three "True" states.

It seems expedient to check whether the relations of Tables 13, 14 and 15 at conjunction correspond to Tables 16, 17 and 18 at disjunction by applying the De

Morgan laws. Such relations exist for the classic propositional logic. This is of great importance when logical operations are between states of different logics.

This is evident by assumption for Tables 13 and 14 and Tables 16 and 17 respectively. For each of the remaining Tables 15 a correspondence will be sought to Tables 18 by checking a control row of each table:

- Negation of the relations of the first row of Table 18 through the De Morgan laws results in

$$\neg(i \vee j) = \neg(i \vee j) \vee \neg(i \vee j) = (F \wedge \neg i) = (F \wedge \neg i) \vee (F \wedge \neg i) = F \wedge \neg i;$$

which corresponds to the relations of the fourth row of Table 18;

Results achieved for the conjunction and disjunction in the S-logic offer a chance some general conclusions to be made:

(1) Tables 13, 14 and 15 thoroughly define the operation "conjunction" between any pair of the six possible states of the set Q of the s_1-logics.
(2) Disjunction between the six states of the set Q of the s_1-logic is thoroughly defined in Tables 16, 17 and 18.
(3) Negation $\neg$ in the s_1-logic is defined by the relations in (18).
(4) Then through the three operations thus defined—conjunction, disjunction, and negation all other relations in the s_1-logic may be defined—implication and equivalence as well as the other formulae, analogic to those of the classic propositional logic and at that keeping in mind constraint (1), equivalent to the negation (21).

A conclusion may be made from the equipollent relations received in the present work

$$\neg i \wedge j = i \vee \neg i \text{ and } i \vee \neg j = i \wedge \neg i$$

that j is not equivalent to i neither is $\neg j$ equivalent to $\neg i$. These relations demonstrate that the state j may be considered as "truer" than i and $\neg j$—as "falser" than $\neg i$. In this sense we may accept that the imaginary logic is "senior" or "more meaningful" than the real one.

4 Relations Between the Real and Two Imaginary Logics

It may be put down from the logical variables $p \in \{i, \neg i\}$ and $q \in = \{j, \neg j\}$ and the results received in the previous three sections:

$$F \wedge i = T; T \wedge i = T; \tag{27}$$

$$T \vee \neg i = F; F \vee \neg i = F; \tag{28}$$

$$\neg i \wedge j = i; i \wedge j = i; \tag{29}$$

$$i \vee \neg j = \neg i; \neg i \vee \neg j = \neg i. \tag{30}$$

The following important results follow from these relations:

Proposition 4.1 *It follows from (27) and (29):*

$$F \wedge \neg i \wedge j = F \wedge (\neg i \wedge j) = F \wedge i = T; \tag{31}$$

$$F \wedge i \wedge j = F \wedge (i \wedge j) = F \wedge i = T; \tag{32}$$

$$T \wedge \neg i \wedge j = T \wedge (\neg i \wedge j) = T \wedge i = T; \tag{33}$$

$$T \wedge i \wedge j = T \wedge (i \wedge j) = T \wedge i = T; \tag{34}$$

Proposition 4.2 *It follows from (28) and (30):*

$$T \vee i \vee \neg j = T \vee (i \vee \neg j) = T \vee \neg i = F; \tag{35}$$

$$T \vee \neg i \vee \neg j = T \vee (\neg i \vee \neg j) = T \vee \neg i = F; \tag{36}$$

$$F \vee i \vee \neg j = F \vee (i \vee \neg j) = T \vee \neg i = F; \tag{37}$$

$$F \vee \neg i \vee \neg j = F \vee (\neg i \vee \neg j) = T \vee \neg i = F. \tag{38}$$

The states corresponding to the ones and zeroes in the logics with indices from $\{i, j, s, s_1, s_2\}$ may be described in the following way:

(a)

$$\text{in the } r \text{ - logic } Q_r = \{T, F\}; \tag{39}$$

(b)

$$\text{in the first imaginary } i \text{ - logic } Q_i = \{i, \neg i\}; \tag{40}$$

(c)

$$\text{in the second imaginary } j \text{ - logic } Q_j = \{j, \neg j\}; \tag{41}$$

(d)

$$\text{in the } s \text{ - logic } Q_s = \{T, F, i, \neg i, (T \vee i), (F \vee \neg i); \tag{42}$$

(e)

$$\text{in the } s_1 \text{ - logic } Q_{s1} = \{i, \neg i, j, \neg j, (i \vee j), (\neg i \wedge \neg j); \tag{43}$$

(f)

$$\text{in the } s_2 \text{ - logic } Q_{s2} = \{Q_r \cup Q_i \cup Q_j \cup Q_s \cup Q_{s1} \cup \{(T \vee i \vee j), (F \wedge \neg i \wedge \neg j) \tag{44}$$

$$Q_{s2} = \{T, F, i, \neg i, j, \neg j, (T \vee i), (F \wedge \neg i), (i \vee j), (F \wedge \neg j), (T \vee i \vee j), (F \wedge \neg i \wedge \neg j)\}. \tag{45}$$

States $\{(T \vee i), (F \vee \neg i)\}$ and $\{(i \vee j), (\neg i \wedge \neg j)\}$ are ones and zeroes, respectively in the s and s_1-logics. This is shown in the previous sections—in (12), (13), (25), and (26). The following relations may be received connecting the ones' and zeroes' states in the separate logics:

Proposition 4.3 *The following relations exist:*

$$(T \vee i) \vee (i \vee j) = T \vee i \vee i \vee j = T \vee i \vee j; \tag{46}$$

$$(F(i \vee j)\neg i) \wedge (\neg i \wedge \neg j) = F \wedge \neg i \wedge \neg i \wedge \neg j = F \wedge \neg i \wedge \neg j); \tag{47}$$

$$(T \vee i) \vee (\neg i \wedge \neg j) = (T \vee i \vee \neg i) \wedge (T \vee i \vee \neg j) = (T \vee i) \wedge F = F \vee (F \wedge i) = T; \tag{48}$$

$$(i \vee j) \wedge (F \wedge \neg i) = (F \wedge \neg i \wedge i) \vee (F \wedge \neg i \wedge j) = (F \wedge \neg i) \vee (F \vee i) = (F \wedge \neg i) \vee T = F; \tag{49}$$

$$(T \vee i \vee j) \vee (F \wedge \neg i \wedge \neg j) = (T \vee i \vee j \vee F) \wedge (T \vee i \vee j \vee \neg i) \wedge (T \vee i \vee j \vee \neg j) = (T \vee i \vee j); \tag{50}$$

$$(T \vee i \vee j) \wedge (F \wedge \neg i \wedge \neg j) = (F \vee \neg i \vee \neg j \wedge T) \vee (F \wedge \neg i \wedge \neg j \wedge i) \wedge (F \wedge \neg i \wedge \neg j \wedge j) = (F \wedge \neg i \wedge \neg j). \tag{51}$$

The relations below show the results of the operations' disjunction and conjunction application to the r, i, and j-logics and s, s_1-logics. They may be comparatively easily obtained on the base of the results described in the present work.

$$(T \vee i) \vee T = (T \vee i); (F \wedge \neg i) \wedge F = (F \wedge \neg i); \tag{52}$$

$$(T \vee i) \vee i = (T \vee i); (F \wedge \neg i) \wedge i = (T \vee i); \tag{53}$$

$$(T \vee i) \vee F = (T \vee i); (F \wedge \neg i) \wedge T = (F \wedge \neg i); \tag{54}$$

$$(T \vee i) \vee \neg i = (T \vee i); (F \wedge \neg i) \wedge \neg i = (F \wedge \neg i); \tag{55}$$

$$(i \vee j) \vee i = (i \vee j); (\neg i \wedge \neg j) \wedge \neg i = (\neg i \wedge \neg j); \tag{56}$$

$$(i \vee j) \vee j = (i \vee j); (\neg i \wedge \neg j) \wedge \neg j = (\neg i \wedge \neg j); \tag{57}$$

$$(i \vee j) \vee \neg i = (i \vee j); (\neg i \wedge \neg j) \wedge i = (\neg i \wedge \neg j); \tag{58}$$

$$(i \vee j) \vee \neg j = (i \vee j); (\neg i \wedge \neg j) \wedge j = (\neg i \wedge \neg j); \tag{59}$$

The results received provide a possibility the truth tables for disjunction and conjunction to be filled in depending the states Q_r, Q_i, Q_j, Q_s, and Q_{s1} of r, i, j-logics and of s, s_1-logics. Table 19 shown corresponds to the disjunction truth table and Table 20—to conjunction. There are unfilled in eight cells in total in both tables. The results received for them are more sophisticated and include formulae with more than three logical variables and due to that they are not put down in these tables.

As it follows from relations (39)–(45) the s_2-logic contains "one" and "zero" of all five logics—r, i, j-logics and s, s_1-logics as well as the states "one" and "zero" characteristic to th s_2-logic only

$$(T \vee i \vee j), (F \wedge \neg i \wedge \neg j). \tag{60}$$

Their properties evolve from the relations (50) and (51).

Table 19 Disjunction

∨	T	F	i	¬i	j	¬j	T∨i	F∧¬i	i∨j	¬i∧¬j
T	T	T	T∨i	F	T∨j	T∨¬j	T∨i	F	T∨i∨j	(¬i∧¬j)∨T
F	T	F	T∨i	F	F∨j	F∨¬j	T∨i	F	F∨i∨j	(¬i∧¬j)∨T
i	T∨i	T∨i	i	i	i∨j	¬j	T∨i	T∨i	i∨j	¬i
¬i	F	F	i	¬i	i∨j	¬j	T∨i	F∧¬i	i∨j	¬i
j	T∨j	F∨j	i∨j	i∨j	j	j	T∨i∨j	(F∧¬i)∨j	i∨j	i∨j
¬j	T∨¬j	F∨¬j	¬j	¬j	j	¬j	F	(F∧¬i)∨¬j	i∨j	¬i∧¬j
T∨i	T∨i	T∨i	T∨i	T∨i	T∨i∨j	F	T∨i	T∨i	T∨i∨j	T
F∧¬i	F	F	T∨i	F∧¬i	(F∧¬i)∨j	(F∧¬i)∨¬j	T∨i	F∧¬i		
i∨j	T∨i∨j	F∨i∨j	i∨j	i∨j	i∨j	i∨j	T∨i∨j		i∨j	i∨j
¬i∧¬j	(¬i∧¬j)∨T	(¬i∧¬j)∨T	¬i	¬i	i∨j	¬i∧¬j	T		i∨j	¬i∧¬j

Table 20 Conjunction

$\wedge$	T	F	i	$\neg i$	j	$\neg j$	$T \vee i$	$F \wedge \neg i$	$i \vee j$	$\neg i \wedge \neg j$
T	T	F	T	$F \wedge \neg i$	$T \wedge j$	$T \wedge \neg j$	T	$F \wedge \neg i$	$(i \vee j) \wedge T$	$T \wedge \neg i \wedge \neg j$
F	F	F	T	$F \wedge \neg i$	$T \wedge j$	$F \wedge \neg j$	T	$F \wedge \neg i$	$(i \vee j) \wedge F$	$F \wedge \neg i \wedge \neg j$
i	T	T	i	$\neg i$	i	$i \wedge \neg j$	$T \vee i$	$F \wedge \neg i$	i	$\neg i \wedge \neg j$
$\neg i$	$F \wedge \neg i$	$F \wedge \neg i$	$\neg i$	$\neg i$	i	$\neg i \wedge \neg j$	$F \wedge \neg i$	$F \wedge \neg i$	i	$\neg i \wedge \neg j$
j	$T \wedge j$	$T \wedge j$	i	i	j	$\neg j$	$T \vee i$	T	$i \vee j$	$\neg i \wedge \neg j$
$\neg j$	$T \wedge \neg j$	$F \wedge \neg j$	$i \wedge \neg j$	$\neg i \wedge \neg j$	$\neg j$	$\neg j$	$(T \vee i) \wedge \neg j$	$F \wedge \neg i \wedge \neg j$	$\neg i \wedge \neg j$	$\neg i \wedge \neg j$
$T \vee i$	T	T	$T \vee i$	$F \wedge \neg i$	$T \vee i$	$(T \vee i) \wedge \neg j$	$T \vee i$	$F \wedge \neg i$		
$F \wedge \neg i$	$F \wedge \neg i$	$F \wedge \neg i$	$F \wedge \neg i$	$F \wedge \neg i$	T	$F \wedge \neg i \wedge \neg j$	$F \wedge \neg i$	$F \wedge \neg i$	F	$F \wedge \neg i \wedge \neg j$
$i \vee j$	$(i \vee j) \wedge T$	$(i \vee j) \wedge F$	i	i	$i \vee j$	$\neg i \wedge \neg j$		F	$i \vee j$	$\neg i \wedge \neg j$
$\neg i \wedge \neg j$	$T \wedge \neg i \wedge \neg j$	$F \wedge \neg i \wedge \neg j$	$\neg i \wedge \neg j$	$\neg i \wedge \neg j$	$\neg i \wedge \neg j$	$\neg i \wedge \neg j$		$F \wedge \neg i \wedge \neg j$	$\neg i \wedge \neg j$	$\neg i \wedge \neg j$

It may be shown that the requirements for commutativity, distributivity, idempotence of disjunction and conjunction are observed in the s_2-logic as well as De Morgan's laws, double negation law, and absorption law and other relations characteristic for the lattice.

In the complicated states like (60) negation is based on De Morgan's laws. For example:

$$\neg(T \vee i \vee j) = \neg(T \vee (i \vee j)) = \neg T \vee (i \vee j) = F \wedge \neg i \wedge \neg j; \qquad (61)$$

$$\neg\neg(T \vee i \vee j) = \neg(F \wedge \neg i \wedge \neg j) = \neg(F \wedge (\neg i \wedge \neg j)) = T \vee \neg(\neg i \wedge \neg j) = T \vee i \vee j. \qquad (62)$$

Last relation confirms the applicability of the double negation law in the s_2-logic.

The requirements ushered in which are also constraints between the r and i-logics and i and j logics

$$F \wedge i = T \text{ and } \neg i \wedge j = i; \qquad (63)$$

and their opposite values

$$T \vee \neg i = F \text{ and } i \vee \neg j = \neg i \qquad (64)$$

lead directly to equalities (31) and (35) and play important role in the relations between the real and imaginary logics.

Proposition 4.4 *The following relations exist in the s_2-logic*

$$F \wedge \neg i \wedge j = T; \qquad (65)$$
$$T \vee i \vee \neg j = F. \qquad (66)$$

The result (65) follows from (63) if in the first equality the state i is substituted by its equivalent $\neg i \wedge j$ from the second equality in (63). In an analogical way if in the first equality from (64) the state $\neg i$ is substituted with its equivalent value $i \vee \neg j$ from the second equality of (64) then (66) will be received.

Movement through (65) and (66) from the r-logic to the i-logic and from there to the j-logic will be denoted as ascending order, and the reverse transitions—as descending order. If some predefined rules when using (65) and (66) are not observed contradictions may appear in the application of associativity of the logical variables in disjunction and conjunction.

Examples that follow demonstrate such possibilities:

Example #1 In the logical expression $(T \vee i \vee \neg j \vee T)$ two possibilities formally exist for defining its value:

(a) $T \vee i \vee \neg j \vee T = T \vee \neg i \vee T = F \vee T = T$;
(b) $T \vee i \vee \neg j \vee T = T \vee \neg i = F$.

The results are different depending on which operation is first executed: idempotence $T \vee T = T$ or $(i \vee \neg j) = \neg i$.

Example #2 Two possibilities exist for defining the logical state of the expression $(T \vee \neg i \vee j)$, namely:

(a) $T \vee \neg i \vee j = F \vee j$;
(b) $T \vee \neg i \vee j = T \vee i$.

Different results are received depending on which of the relations (63) and (64) is first used.

The contradictions pointed out in the two examples could be avoided if the following two step rule is observed when realizing operations disjunction and conjunction and relations (63) and (64).

Rule A:

First step: All possible and necessary disjunctive and conjunctive logical operations are performed in the framework of the r, i, and j-logics in ascending order of the variables $\{T, F\}$ to the logical variables $\{i \vee \neg i\}$ and at the end—with variables $\{j \vee \neg j\}$. Where the formulae contain variables of the r, i, and j-logics the necessary logical operations are carried out mixed between those two formulae and variables. In no way at the realization of the first step formulae (63) and (64) should be used. *Second step*: After exhausting all possible cases of the first step relations (63) and (64) should be applied in descending order—first of all between the variables $\{i, \neg i, j, \neg j\}$ and then between $\{T, F, i, \neg i\}$.

Rule A above described helps avoiding the contradictions at applying of the associativity of the logical variables for disjunction and conjunction.

It follows from the results exposed that the simultaneous considering of the real propositional logic and the two imaginary logics provide a possibility this structure to be considered as a multiple-valued logic with 12 states in general—6 for r, i, j-logics and so many for the complex s, s_1, s_2-logics. At that each of these logics is separately a two-valued logic.

The introduction of a second imaginary two-valued j-logic increases twice the number of states for the initial r, i, s-logics.

Applicability of the complex logical structures being investigated may be sought in various areas of science and practice—linguistics, social processes, technological systems, etc.

References

1. Kleene, S.C., Mathematical Logic, J. Wiley & Sons, N.Y., 1967.
2. Kowalski, R., Logic for Problem Solving, Elsevier North Holland Inc., 1979.
3. Sgurev, V., An Essay on Complex Valued Propositional Logic, Artificial intelligence and Decision Making, №2, 2014, pp. 95–101, Institute of System Analysis RAS (Институт системного анализа РАН), http://www.aidt.ru, ISSN 2071-8594.

4. D.E. Tamir, A. Kandel, Axiomatic Theory of Complex Fuzzy Logic and Complex Fuzzy Classes, Int. J. of Computers, Communications & Control, ISSN 1841-9836, E-ISSN 1841-9844, Vol. VI (2011), No. 3 (September), pp. 562–576.
5. Richard G. Shoup, A Complex Logic for Computation with Simple Interpretations for Physics, Interval Research Palo Alto, CA 94304, http://www.boundarymath.org/papers/CompLogic.pdf.
6. Igor Aizenberg, Complex-Valued Neural Networks with Multi-Valued Neurons, Springer, 2011.
7. Louis H. Kauffman, Virtual Logic — The Flagg Resolution, Cybernetics & Human Knowing, Vol. 6, no. 1, 1999, pp. 87–96.
8. Sgurev, V., An Approach to Constructing a Complex Propositional Logic, Comptes Rendus Acad. Bulg. Sci., Tome 66, № 11, pp. 1623-1632, ISSN 1310_1331, "Marin Drinov" Academic Publishing House, Sofia, 2013.
9. Sgurev, V., On Way to the Complex Propositional Logic, Int. Conf. "Automatics and Informatics – 2013", 3–7 Oct., Sofia, Bulgaria. 2013.
10. Johnson, B., Topics in Universal Algebra, Springer-Verlag, N.Y., 1972.
11. Ionov, A.S., G.A. Petrov, Quaternion Logic as a Base of New Computational Logic, http://zhurnal.ape.relarn.ru/articles/2007/047.pdf (in Russian).
12. Hung T. Nguyen, Vladik Kreinovich, Valery Shekhter, On the Possibility of Using Complex Values in Fuzzy Logic For Representing Inconsistencies, http://www.cs.utep.edu/vladik/1996/tr96-7b.pdf.
13. Chris Lucas, A Logic of Complex Values, Proceedings of the First International Conference on Neutrosophy, Neutrosophic Logic, Set, Probability and Statistics, 1–3 December 2001, University of New Mexico, pp. 121–138. ISBN 1-931233-55.
14. V. Sgurev, Logical Operations and Inference in the Complex s-logic, Chapter Innovative Issues in Intelligent SystemsVolume 623 of the series Studies in Computational Intelligence pp 141–160, https://doi.org/10.1007/978-3-319-27267-2_5, © Springer International Publishing AG. Part of Springer Nature.
15. Mendelson, E., Introduction to Mathematical Logic, D.von Nostrand Com. Inc., Princeton, 1975.

Generalized Net Model of the Processes in a Center of Transfusion Haematology

Nikolay Andreev, Evdokia Sotirova, Anthony Shannon and K T Atanassov

Abstract The proceeses in a center of transfusion haematology—receiving of person's blood, obtaining of its fresh frozen plasma, erythrocytes and thrombocytes, its testing for transmissible diseases (HIV, HBV, HCV, Wass) and evaluation of blood group and Rh, and antibodies screening—are described by a generalized net.

Keywords Generalized net · Transfusion haematology

1 Introduction

As it is mentioned in [5] on 27 January 2003 the European Union adopted Directive 2002/98/EC on setting standards of quality and safety for the collection, testing, processing, storage and distribution of human blood and blood components.

In the present research, following the instructions from [5], we describe the process of collecting, testing, separating, evaluating, keeping and distributing of human blood. As a tool for description of this process, we use the apparatus of

N. Andreev
National Center of Transfusion Haematology, 112, Bratya Miladinovi Street, 1202 Sofia, Bulgaria
e-mail: imuno_chem@abv.bg

E. Sotirova · K T Atanassov
Prof. Asen Zlatarov University, 8010 Bourgas, Bulgaria
e-mail: esotirova@btu.bg

K T Atanassov (✉)
Department of Bioinformatics and Mathematical Modelling, Institute of Biophysics and Biomedical Engineering, Bulgarian Academy of Sciences, Acad. G. Bonchev Street, Bl. 105, 1113 Sofia, Bulgaria
e-mail: krat@bas.bg

A. Shannon
Honorary Fellow, Warrane College, The University of New South Wales, 2033 Kensington, NSW, Australia
e-mail: t.shannon@warrane.unsw.edu.au

© Springer International Publishing AG, part of Springer Nature 2019
M Hadjiski and K T Atanassov (eds.), *Intuitionistic Fuzziness and Other Intelligent Theories and Their Applications*, Studies in Computational Intelligence 757,
https://doi.org/10.1007/978-3-319-78931-6_8

the Generalized Nets (GNs; see [1–3]), because they give a suitable instrument for describing the above mentioned processes in their dynamic development. In Sect. 2, short remarks on GNs are given and in Sect. 3 the GN-model is described.

We must immediately mention, that in the present paper we use a simpler form of a GN, that in [2, 3] is called a reduced GN.

2 Short Remarks on the Theory of the Generalized Nets

GNs are an extension as of the standard Petri nets, as well as of the rest of their extensions and modifications. GNs are defined in a way that is principally different from the ways of defining the other types of Petri nets (see [2, 3]).

When some of the GN-components are omitted, the GN is called a reduced GN. For the needs of the model below, we describe the modelleed process as a reduced GNs.

Formally, every transition (see Fig. 1) is described by a seven-tuple, but for our aims, we use its following reduced form:

$$Z = \langle L', L'', r \rangle,$$

where:

- L' and L'' are finite, non-empty sets of places (the transition's input and output places, respectively); for the transition in Fig. 1 these are $L' = \{l'_1, l'_2, \ldots, l'_m\}$ and $L'' = \{l''_1, l''_2, \ldots, l''_n\}$;

Fig. 1 The form of a GN-transition

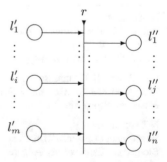

- r is the transition's *condition* determining which tokens will pass (or *transfer*) from the transition's inputs to its outputs; it has the form of an Index Matrix (IM; see [4]):

$$r = \begin{array}{c|c} & l''_1 \dots l''_j \dots l''_n \\ \hline l'_1 & \\ \vdots & r_{i,j} \\ l'_m & \end{array} \quad ;$$

$r_{i,j}$ is the predicate that corresponds to the ith input and jth output place ($1 \le i \le m$, $1 \le j \le n$). When its truth value is "*true*", a token from the ith input place transfers to the jth output place; otherwise, this is not possible.

In general, the GN is defined as ordered four-tuple, but in the present case, it has the form

$$E = \langle A, K, X, \Phi \rangle,$$

where:

- A is a set of transitions;
- K is the set of the GN's tokens.
- X is the set of all initial characteristics which the tokens can obtain on entering the net;
- Φ is the characteristic function that assigns new characteristics to every token when it makes the transfer from an input to an output place of a given transition.

Operations, relations and operators are defined over GNs (see [2, 3]).

3 Generalized Net Model

The present GN (see Fig. 2) contains 15 transitions, 38 places and 8 types of tokens. Tokens $\beta_1, \beta_2, \dots$ enter place l_1 with the initial characteristic

"identifier, name address and telephone of i-th person; time-moment

(date, hour, minute), quantity of the blood, duration of blood donation",

where $i = 1, 2, \dots$. For brevity, this index will be omitted, but when we discuss token β- γ-, etc. tokens, we will have in mind β_i- γ_i-, etc. tokens.

$$Z_1 = \langle \{l_1\}, \{l_2, l_3, l_4\}, \begin{array}{c|ccc} & l_2 & l_3 & l_4 \\ \hline l_1 & true & true & true \end{array} \rangle.$$

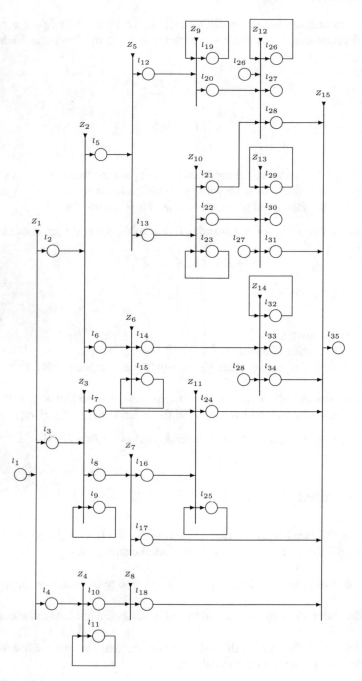

Fig. 2 GN-model

Token β splits to three tokens β, γ and δ that enter places l_2, l_3, l_4, respectively, with characteristics

"blood for Fresh Frozen Plasma (FFP), thrombocytes and erythrocytes"

for token β,

"quantity of the blood for tests for transmissible diseases (HIV, HBV,

HCV, Wass)"

for token γ,

"quantity of the blood for evaluation of blood group, Rh and antibodies

screenings"

for token δ.

$$Z_2 = \langle \{l_2\}, \{l_5, l_6\}, \frac{\begin{array}{c|cc} & l_5 & l_6 \\ \hline l_2 & true & true \end{array}}{} \rangle.$$

Token β from place l_2 splits to two tokens β and ε that enter places l_5 and l_6, respectively, with characteristics

"plasma, quantity"

for token β,

"erythrocytes, quantity"

for token ε.

$$Z_3 = \langle \{l_3, l_9\}, \{l_7, l_8, l_9\}, \begin{array}{c|ccc} & l_7 & l_8 & l_9 \\ \hline l_3 & false & false & true \\ l_9 & W_{9,7} & W_{9,8} & false \end{array} \rangle,$$

where
$W_{9,7} = $ "The results of the transmissible diseases tests are negative",
$W_{9,8} = \neg W_{9,7}$,
i.e., there is at least one positive test for transmissible disease. Here, $\neg P$ denotes the negation of predicate P.

Token γ from place l_3 enters place l_9 with characteristics

"results of the transmissible diseases tests".

On the next time step, token γ enters one of the other output places with characteristics "the blood of the i-th person is suitable for use"

in place l_7 and

"the blood of the i-th person must be tested again"

in place l_8.

$$Z_4 = \langle \{l_4, l_{11}\}, \{l_{10}, l_{11}\}, \begin{array}{c|cc} & l_{10} & l_{11} \\ \hline l_4 & false & true \\ l_9 & true & false \end{array} \rangle.$$

Token δ from place l_4 enters place l_{11} with a characteristic

"results of the evaluation of blood group and Rh".

On the next time step, token δ enters place l_{10} with a characteristic

"results of the evaluation of antibodies screenings".

$$Z_5 = \langle \{l_5\}, \{l_{12}, l_{13}\}, \begin{array}{c|cc} & l_5 & l_6 \\ \hline l_5 & true & true \end{array} \rangle.$$

Token β splits to two tokens φ and τ that enter places l_{12} and l_{13}, respectively, with characteristics

"FFP without thrombocytes, quantity, time-moment

(date, hour, minute)"

for token φ,

"Fresh Frozen Plasma (FFP) rich in thrombocytes, quantity, time-moment

(date, hour, minute)"

for token τ.

$$Z_6 = \langle \{l_6, l_{15}\}, \{l_{14}, l_{15}\}, \begin{array}{c|cc} & l_{14} & l_{15} \\ \hline l_6 & false & true \\ l_{15} & W_{15,14} & W_{15,15} \end{array} \rangle,$$

where
$W_{15,14} = $ "the test results are ready",
$W_{15,15} = \neg W_{15,14}$,
   From place l_6, token ε enters place l_{15} with a characteristic

"safe-keeping deadline of the erythrocytes of i-th person".

When $W_{15,15} = true$, it enters place l_{14} with a characteristic

"there is an evaluation for the i-th person's blood".

$$Z_7 = \langle \{l_8\}, \{l_{16}, l_{17}\}, \frac{\begin{array}{c|cc} & l_{16} & l_{17} \\ \hline l_8 & W_{8,16} & W_{8,17} \end{array}}{} \rangle,$$

where:
$W_{8,16} = $ "The results of the second tests for transmissible diseases are negative",
$W_{8,17} = \neg W_{8,16}$.
Token γ from place l_8 enters place l_{16} with characteristics

"the blood of the i-th person is suitable for use"

and in place l_{17} with characteristics

"the blood of the i-th person must be disposed of".

$$Z_8 = \langle \{l_{10}\}, \{l_{18}\}, \frac{\begin{array}{c|c} & l_{18} \\ \hline l_{10} & true \end{array}}{} \rangle.$$

Token δ from place l_{10} enters places l_{18} with a characteristic

"the blood of the i-th person's must be disposed of".

$$Z_9 = \langle \{l_{12}, l_{19}\}, \{l_{19}, l_{20}\}, \frac{\begin{array}{c|cc} & l_{19} & l_{20} \\ \hline l_{12} & true & false \\ l_{19} & W_{20,20} & W_{20,21} \end{array}}{} \rangle,$$

where
$W_{19,19} = $ "there are no tests results for the ith person",
$W_{19,20} = \neg W_{19,19}$.
Token φ from place l_{12} enters place l_{19} and from this place, it enters place l_{20} with a characteristic

"duration of the mandatory quarantine period".

$$Z_{10} = \langle \{l_{13}, l_{23}\}, \{l_{21}, l_{22}, l_{23}\}, \frac{\begin{array}{c|ccc} & l_{21} & l_{22} & l_{23} \\ \hline l_{13} & false & false & true \\ l_{23} & W_{23,21} & W_{23,22} & W_{23,23} \end{array}}{} \rangle,$$

where
$W_{23,23}$ = "there are no tests results for the ith person",
$W_{23,21} = W_{23,22} = \neg W_{23,23}$.

Token τ from place l_{13} enters place l_{23} without any characteristic. On in a next moment, when $W_{23,23} = false$, i.e., $W_{23,21} = W_{23,22} = true$, it splits to two tokens: φ and τ. The first of them enters place l_{21} with a characteristic

"FFP without thrombocytes, quantity, time-moment (date, hour, minute)".

The second token enters place l_{22} with a characteristic

"thrombocytes, quantity, time-moment (date, hour, minute)".

$$Z_{11} = \langle \{l_7, l_{16}, l_{25}\}, \{l_{24}, l_{25}\}, \begin{array}{c|cc} & l_{24} & l_{25} \\ \hline l_7 & false & true \\ l_{16} & false & true \\ l_{25} & W_{25,24} & W_{25,25} \end{array} \rangle,$$

where
$W_{25,24}$ = "the preservation period of the tests results has finished",
$W_{25,25} = \neg W_{25,24}$.

Token γ from place l_7 or from place l_{16} enters place l_{25} with a characteristic

"the i-th person's tests results, time-moment (date, hour, minute)"

and it enters places l_{24} with a characteristic

"the safekeeping deadline of the i-th person's tests results has expired".

Token ρ_φ enters place l_{26} with the initial characteristic

"request from hospital (name, address) for FFP (quantity)".

$$Z_{12} = \langle \{l_{20}, l_{21}, l_{26}, l_{29}\}, \{l_{29}, l_{30}, l_{31}\}, \begin{array}{c|ccc} & l_{29} & l_{30} & l_{31} \\ \hline l_{20} & true & false & false \\ l_{21} & true & false & false \\ l_{26} & false & true & false \\ l_{29} & W_{29,29} & W_{29,30} & W_{29,31} \end{array} \rangle,$$

where
$W_{29,29}$ = "the FFP is in the safekeeping deadline (2 years) and there has not been a request for it" or "transmissible diseases tests of the ith person are positive" or "the blood of the ith person contains antibodies",
$W_{29,30}$ = "the FFP is in the term of limitation and there is a request for it",
$W_{29,31}$ = "the FFP is already outside the term of limitation".

Token φ from place l_{20} or place l_{21} enters place l_{29} with the characteristics

"information for i-th person's blood group and Rh; the FFP must be preserved

at $-25\,°C$".

On a next time step, when $W_{29,30} = true$, token φ enters place l_{30} together with token ρ_{φ} and the two tokens unite with characteristic

"name of the person who will obtain the FFP; the identifier(s)

of the bank(s) with FFP".

When $W_{29,31} = true$, token φ enters place l_{31} without any characteristic. Token ρ_{τ} enters place l_{27} with the initial characteristic

"request from hospital (name, address) for thrombocytes (quantity)".

$$Z_{13} = \langle \{l_{22}, l_{27}, l_{32}\}, \{l_{32}, l_{33}, l_{34}\}, \begin{array}{c|ccc} & l_{32} & l_{33} & l_{34} \\ \hline l_{22} & true & false & false \\ l_{27} & false & true & false \\ l_{32} & W_{32,32} & W_{32,33} & W_{32,34} \end{array} \rangle,$$

where
$W_{32,32} =$ "the thrombocytes are in the term of limitation (5 days) and there is not a request for them" or "transmissive diseases tests of the ith person are positive" or "the blood of the ith person contains antibodies",
$W_{32,33} =$ "the thrombocytes are in the term of limitation and there is a request for them",
$W_{32,34} =$ "the thrombocytes are already outside the term of limitation".
   Token τ from place l_{22} enters places l_{32} with a characteristics

"information for i-th person's blood group and Rh; the thrombocytes must be

preserved at $+22\,°C$".

On a next time step, when $W_{32,33} = true$, token τ enters place l_{33} together with token ρ_{τ} and the two tokens unite with a characteristic

"name of the person who will obtain the thrombocytes;

identifier(s) of the bank(s) with thrombocytes".

When $W_{32,34} = true$, token τ enters place l_{34} without any characteristic. Token ρ_{ε} enters place l_{28} with the initial characteristic "request from hospital (name, address) for erythrocytes (quantity)".

$$Z_{14} = \langle \{l_{14}, l_{28}, l_{35}\}, \{l_{35}, l_{36}, l_{37}\}, \begin{array}{c|ccc} & l_{35} & l_{36} & l_{37} \\ \hline l_{14} & true & false & false \\ l_{28} & false & true & false \\ l_{35} & W_{35,35} & W_{35,36} & W_{35,37} \end{array} \rangle,$$

where

$W_{35,35} =$ "the erythrocytes are in the safekeeping deadline (35–42 days) and there is not a request for them" or "transmissible diseases tests of the ith person are positive" or "the blood of the ith person contains antibodies",

$W_{35,36} =$ "the erythrocytes are in the term of limitation and there is a request for them",

$W_{35,37} =$ "the erythrocytes are already outside the term of limitation".

Token ε from place l_{14} enters places l_{35} with characteristics

"information for i-th person's blood group and Rh; the erythrocytes must be

preserved at temperature from $+4\,°$to $+8\,°$C.

On a next time step, when $W_{35,36} = true$, token ε enters place l_{36} together with token ρ_ε and the two tokens unite with a characteristic

"name of the person who will obtain the erythrocytes; identifier(s)

of the bank(s) with thrombocytes".

When $W_{35,37} = true$, token ε enters place l_{37} without any characteristic.

$$Z_{15} = \langle \{l_{17}, l_{18}, l_{24}, l_{31}, l_{34}, l_{37}\}, \{l_{38}\}, \begin{array}{c|c} & l_{38} \\ \hline l_{17} & true \\ l_{18} & true \\ l_{24} & true \\ l_{31} & true \\ l_{34} & true \\ l_{37} & true \end{array} \rangle.$$

Each of the tokens from the input places of the transition enters place l_{38} with characteristic

"information about the disposed biological material".

4 Conclusion

The so constructed GN-model can be used for several aims. First, it describes the development of the process of manipulation with human blood. Second, it can be used for searching of possibility for optimization of these processes in time. Third, it can be used for control of these processes in real-time, especially in the case, when it is used as a sub-GN in a GN, describing the separate blood-collection-centers and hospital, that need the blood. So, in future, we will construct such more extended GNs of which the present GN will be a part.

References

1. Alexieva, J., Choy, E., E. Koycheva. Review and bibloigraphy on generalized nets theory and applications. In: A Survey of Generalized Nets (E. Choy, M. Krawczak, A. Shannon and E. Szmidt, Eds.), Raffles KvB Monograph No. 10, 2007, 207–301.
2. Atanassov, K. Generalized Nets. World Scientific. Singapore, London, 1991.
3. Atanassov, K. On Generalized Nets Theory. Prof. M. Drinov Academic Publ. House, Sofia, 2007.
4. Atanassov, K. Index Matrices: Towards an Augmented Matrix Calculus, Springer, Cham, 2014.
5. Guide to the Preparation, Use and Quality Assurance of Blood Components. Recommendation No. R (95) 15, 19th Edition (Dr S. Keitel, Ed.), European Directorate for the Quality of Medicines & HealthCare, Council of Europe, Strasbourg, 2017.

Image to Sound Encryption Using a Self-organizing Map Neural Network

Todor Petkov and Sotir Sotirov

Abstract This paper describes the process of encrypting image in a sound using artificial neural network. In order to achieve it the process is divided into several steps where each of the steps is described with a generalized net. The main goal is to send an image which is encrypted into a sound between two persons and if a wrong person receives it he will not be permitted to see the image. The neural network is divided into 5 clusters where each cluster responds to areas where the image has to be encrypted. When the procedure ends a random sound is applied to the network for testing and depending on which cluster it enters the necessarily areas are taken and the image is applied on them.

Keywords Encryption · Generalized net · Neural networks · Image processing

1 Introduction

An artificial neural network is an information-processing system that has certain performance characteristics in common with biological neural networks [1]. The biological neural network consists of small cells called neurons, each neuron consists of cell body, axon, synapses and dendrites in which the chemical signals proceed through them (Fig. 1) [1, 2].

Encryption is the process of encoding messages or information in such a way that only authorized parties can read it [3]. Decryption is the process of decoding data that has been encrypted into a secret format [3]. There are two main types of encryption—one with symmetric and one with asymmetric key.

The same key is used for encrypting and decrypting in algorithms that use symmetric key (Fig. 2) and an example of such algorithm is DES (data encryption standard).

T. Petkov (✉) · S. Sotirov
Laboratory of Intelligent Systems, University "Prof. Dr. Assen Zlatarov", Burgas, Bulgaria
e-mail: todor_petkov@btu.bg

S. Sotirov
e-mail: ssotirov@btu.bg

© Springer International Publishing AG, part of Springer Nature 2019   135
M Hadjiski and K T Atanassov (eds.), *Intuitionistic Fuzziness and Other Intelligent Theories and Their Applications*, Studies in Computational Intelligence 757,
https://doi.org/10.1007/978-3-319-78931-6_9

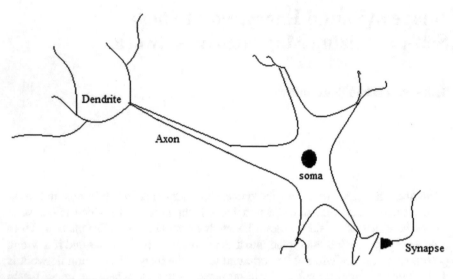

Fig. 1 Model of biological neuron

Fig. 2 Encryption with symmetric key

In an asymmetric key encryption, anyone can encrypt the message using the public key, but only the holder of the paired private key can decrypt the message [3].

In physics, sound is a vibration that propagates as a typically audible mechanical wave of pressure and displacement, through a medium such as air or water. In physiology and psychology, sound is the reception of such waves and their perception by the brain [4, 5]. Each sound wave can be big or small: big sound waves have what's called a high amplitude or intensity and we hear them as louder sounds. Apart from amplitude, another thing worth noting about sound waves is their pitch, also called their frequency. The frequency is simply the number of waves something produces in one second (Fig. 3).

An image can be defined as a two-dimensional function, $f(x, y)$, where x and y are spatial coordinates of the image [6]. The elements of the image that have particular location are called picture elements (pixels). Pixel is the term that is used most widely for the elements of digital image. The color of the pixel can be defined from the three basic colors that human eye can see: red (R), green (G) and blue (B). Each image can be represented as three layers with those three colors: the first layer corresponds to the red, the second to the green and the third to the blue color (Fig. 4).

Pixels from the image are suitable for evaluation for different purposes such as encryption [7–10].

Fig. 3 Different sound waves

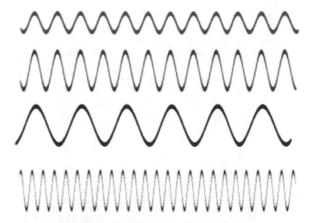

Fig. 4 Pixel representation of the image

Self Organizing Map (SOM) (Fig. 5) is an unsupervised neural network (NN) introduced from Teuvo Kohonen. SOM is also known as topology-preserving maps, it assumes topological structure among the cluster units [1, 11]. During the self-organization process the cluster unit (neuron) whose weight vector matches most closely to the input pattern is chosen as a winner. Next, the weight vectors for all neurons within a certain neighborhood of the winner are updated using the Kohonen rule.

Learning algorithm according to [12] is:

Step 0   Initialize weights $W_{i,j}$
          Set topological neighborhood parameters
          Set learning rate parameters
Step 1   While stopping conditions is false, do Steps 2–8.
Step 2   For each input vector x, do Steps 3–5.
Setp 3   For each j, compute:

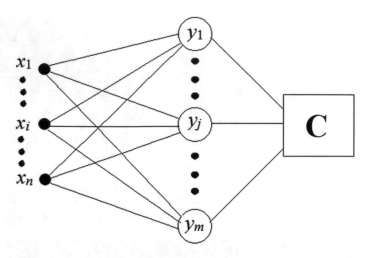

Fig. 5 Structure of Self Organizing Map NN

$$D(j) = \sum_{1}^{n} \left(w_{ij} - x_i\right)^2$$

Step 4 Find index J such that $D(J)$ is a minimum.
Step 5 For all units j within a specified neighborhood of J, and for all i:

$$w_{ij}(new) = w_{ij}(old) + \alpha(x_i - w_{ij}(old))$$

Step 6 Update learning rate.
Step 7 Reduce radius of topological neighborhood at specified times
Step 8 Test stopping condition

2 GN-Model

Initially the following tokens enter the Generalized Net (GN) [13].

In place L_1—"α token with characteristic input image for encryption".

In place L_2—"β token with characteristic input parameters for processing of the image".

In place L_3—"γ token with characteristic input parameters for learning of the S.O.M.".

In place L_4—"δ token with characteristic input vectors for training the S.O.M.".

In place L_{10}—"η token with characteristic sound for encrypting the image".

The GN—(Fig. 6) is introduced by the following set of transitions:

$A = \{Z_1, Z_2, Z_3, Z_4, Z_5, Z_6\}$,

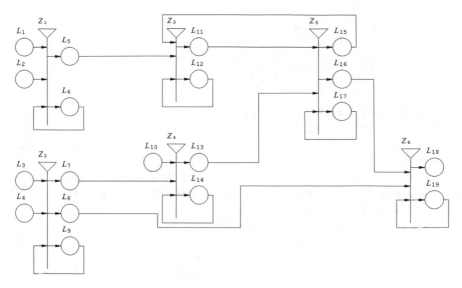

Fig. 6 GN – model of encrypting image in a sound

$Z_1 =$ "Image processing";

$Z_2 =$ "Learning of the S.O.M. NN";

$Z_3 =$ "Transformation of a component from the image";

$Z_4 =$ "Verification of clusters";

$Z_5 =$ "Encrypting the image";

$Z_6 =$ "Decrypting the image"

GN consists of six transitions by the following descriptions:

$Z_1 = \langle \{L_1, L_2, L_6\}, \{L_5, L_6\}, R_1 \vee (L_1, L_2, L_6) \rangle$

$$R_1 = \frac{\begin{array}{c|cc} & L_5 & L_6 \end{array}}{\begin{array}{c|cc} L_1 & false & true \\ L_2 & false & true \\ L_6 & W_{6,5} & W_{6,6} \end{array}}$$

$W_{6,5} =$ "The image is processed",

$W_{6,6} =$ "There are tokens α and β".

Here and below, objects R_i $(i = 1, 2, ..., 6)$ are index matrices [14, 15].

Token α that enters place L_6 from place L_1 does not obtain new characteristic.

Token α that enters place L_6 from place L_6 obtains characteristic

$x_{cu}^{\alpha'} =$ "Processing of the image".

Token α' that enters place L_5 from place L_6 obtains characteristic $x_{cu}^{\alpha''} =$ "Processing image".

$Z_2 = \langle \{L_3, L_4, L_9\}, \{L_7, L_8, L_9\}, R_2 \vee (L_3, L_4, L_9) \rangle$

$$R_2 = \begin{array}{c|ccc} & L_7 & L_8 & L_9 \\ \hline L_3 & false & false & true \\ L_4 & false & false & true \\ L_9 & W_{9,7} & W_{9,8} & W_{9,9} \end{array}$$

$W_{9,7} = W_{9,8} = $ "The S.O.M. NN is trained",
$W_{9,9} = $ "There are γ and δ tokens".

Token γ from place L_3 and token γ from place L_4 unite in place L_9 and obtain characteristic $x_{cu}^{\varepsilon} = $ "Training the S.O.M. NN".

Token ε that enters place L_9 from place L_9 does not obtain new characteristic.

Token ε splits and enters places L_7 and L_8 from place L_9 obtains characteristic $x_{cu}^{\varepsilon'} = $ "Trained S.O.M. NN".

$Z_3 = \langle \{L_5, L_{12}, L_{15}\}, \{L_{11}, L_{12}\}, R_3 \vee (L_5 \wedge (L_{12}, L_{15})) \rangle$

$$R_3 = \begin{array}{c|cc} & L_{11} & L_{12} \\ \hline L_5 & false & true \\ L_{12} & W_{12,11} & W_{12,12} \\ L_{15} & false & true \end{array}$$

$W_{12,11} = $ "The image component is extracted".
$W_{12,12} = $ "There are tokens α'' and κ'".

Token α'' from place L_5 that enters place L_{12} does not obtain new characteristic.

Token α'' from place L_{12} that enters place L_{12} obtain characteristic
$x_{cu}^{\alpha'''} = $ "Extracting of the image component".

Token α''' from place L_{12} unites with token κ' from place L_{15} and obtain characteristic $x_{cu}^{\varphi} = $ "Image component for encrypting(i) and $i = i + 1$".

Token φ from place L_{12} that enters place L_{11} does not obtain new characteristic.

$Z_4 = \langle \{L_7, L_{10}, L_{14}\}, \{L_{13}, L_{14}\}, R_4 \vee (L_7, L_{10}, L_{14}) \rangle$

$$R_4 = \begin{array}{c|cc} & L_{13} & L_{14} \\ \hline L_7 & false & true \\ L_{10} & false & true \\ L_{14} & W_{14,13} & W_{14,14} \end{array}$$

$W_{14,13} = $ "The clusters are verified",
$W_{14,14} = $ "There are ε' and η tokens".

Token ε' from place L_7 unites with token η from place L_{10} in place L_{14} and obtain characteristic $x_{cu}^{\lambda} = $ "The sound component is tested".

Token λ from place L_{14} that enters place L_{13} obtains characteristic $x_{cu}^{\lambda'} = $ "The cluster is determined".

Token λ from place L_{14} that enters place L_{14} does not obtain characteristic.

$Z_5 = \langle \{L_{11}, L_{13}, L_{17}\}, \{L_{15}, L_{16}, L_{17}\}, R_5 \vee (L_{17} \wedge (L_{11}, L_{13})) \rangle$

$$R_5 = \frac{\begin{array}{c|ccc} & L_{15} & L_{16} & L_{17} \\ \hline L_{11} & false & false & true \\ L_{13} & false & false & true \\ L_{17} & W_{17,15} & W_{17,16} & W_{17,17} \end{array}}{}$$

$W_{17,15} =$ "The component is encrypted",
$W_{17,16} =$ "The image is encrypted".
Token φ from place L_{11} unites with token λ' from place L_{13} in place L_{15} with characteristic
$x_{cu}^{\kappa} =$ "Encrypted image in sound".
Token κ from place L_{17} that enters place L_{15} obtains characteristic
$x_{cu}^{\kappa'} =$ "Request for next image component".
Token κ from place L_{17} that enters place L_{16} does not obtain new characteristic.
$Z_6 = \langle \{L_8, L_{16}, L_{19}\}, \{L_{18}, L_{19}\}, R_6 \vee (L_{19} \wedge (L_8, L_{16})) \rangle$

$$R_6 = \frac{\begin{array}{c|cc} & L_{18} & L_{19} \\ \hline L_8 & false & true \\ L_{16} & false & true \\ L_{19} & W_{19,18} & W_{19,19} \end{array}}{},$$

$W_{19,18} =$ "The image is decrypted",
$W_{19,19} =$ "There are ε' and κ'' tokens"
Tokens ε' and κ'' from places L_9 and L_{16} unites and enters place L_{18} with characteristic $x_{cu}^{\mu} -$ "Visualization of the encrypted image".

3 Conclusion

In this paper is purposed a method for encrypting image in a sound using neural network. The type of neural network used for the aim is Self Organizing Map. The training set is taken from a random sound where the extracted vector is with size of 1 by 100 elements. When the procedure is of training is done another sound is taken in order to test and determine where to encrypt the image. In this case when the sound is sent to the receiver, after its testing, he will receive the correct areas to take the image components. If unauthorized person receives the information without a trained SOM he will not be able to see an image. The process of image encryption is represented with the apparatus of generalized net where it is useful in cases where there are parallel processes. In this case the Kohonen Self Organizing Map is symmetrical key for encrypting and decrypting the image.

References

1. Beale M., Demuth H., Hagan M., Neural Network Design, PWS Publishing Company, 1996.
2. Jain, A., Jianchang M., Artificial Neural Networks: A Tutorial, IEEE, 1996.
3. Goldreich, O., Foundations of Cryptography, Volume II Basic A0pplications, Cambridge University Press, 2004.
4. Chen, G., Vijay Parsa, Objective Speech Quality Evaluation Using an Adaptive Neuro-Fuzzy Network, Speech, Audio Image and Biomedical Signal Processing using Neural Networks, Springer [97 – 116].
5. S., Andreas, T. Painter, V. Atti, Audio Signal Processing and Coding, John Wiley & Sons, 2006.
6. Gonzales, R., R. Woods, Digital Image Processing, Third Edition, Pearson Education, Inc., 2008.
7. Kumar, M., Vaish, A. Encryption of color images using MSVD in DCST domain (2017) Opt Lasers Eng, 88, pp. 51–59.
8. Rao, P., Audio Signal Processing Speech, Audio Image and Biomedical Signal Processing using Neural Networks, Springer [169 – 189].
9. Refreiger, P., Javidi, B.Optical image encryption based on input plane and Fourier plane random encoding (1995) Opt Lett, 20, pp. 767–769.
10. Li, X., Zhao, D. Optical color image encryption with redefined fractional Hartley transform (2010) Optik, 121, pp. 673–677.
11. Vesanto, J., E. Alhoniemi, Clustering of the Self-Organizing Map, IEEE Transactions on Neural Networks, vol. 11, NO. 3, 2000.
12. Fausett, L., Fundamentals of Neural Networks, Architecture algorithms and applications, 1993.
13. Atanassov K. Generalized nets. World Scientific, Singapore, New Jersey, London, 1991.
14. Atanassov K., Generalized index matrices, Comptes rendus de l'Academie Bulgare des Sciences, vol.40, 1987, No.11, 15–18.
15. Atanassov, K., Index Matrices: Towards an Augmented Matrix Calculus, Springer, Cham, 2014.

On Different Algorithms for InterCriteria Relations Calculation

Olympia Roeva, Peter Vassilev, Nikolay Ikonomov, Maria Angelova, Jun Su and Tania Pencheva

Abstract Contemporary InterCriteria analysis (ICrA) approach for searching of existing or unknown correlations between multiple objects against multiple criteria is applied here. Altogether five different algorithms for InterCriteria relations calculation have been examined to render the influence of the genetic algorithm parameters on the algorithm performance. Two cases, i.e. the model parameter identification of *E. coli* and *S. cerevisiae* fed-batch fermentation processes, are considered. In this investigation μ-biased, Balanced, ν-biased, Unbiased, as well as the newly elaborated and proposed here Weighted algorithm have been consequently applied and thoroughly examined. The obtained results for considered here two Case studies have been compared showing that the most reliable algorithm is the μ-biased one.

Keywords InterCriteria analysis · Intuitionistic fuzzy sets · Genetic algorithms
Parameter identification · Fermentation processes · *E. coli* · *S. cerevisiae*

All authors have contributed equally to this work.

O. Roeva (✉) · P. Vassilev · M. Angelova · T. Pencheva
Institute of Biophysics and Biomedical Engineering,Bulgarian Academy of Sciences,
Sofia, Bulgaria
e-mail: olympia@biomed.bas.bg
P. Vassilev
e-mail: peter.vassilev@gmail.com

N. Ikonomov
Institute of Mathematics and Informatics,Bulgarian Academy of Sciences,Sofia, Bulgaria
e-mail: nikonomov@math.bas.bg
M. Angelova
e-mail: maria.angelova@biomed.bas.bg

J. Su
School of Computer Science,Hubei University of Technology, Hubei, China
T. Pencheva
e-mail: tania.pencheva@biomed.bas.bg

© Springer International Publishing AG, part of Springer Nature 2019
M Hadjiski and K T Atanassov (eds.), *Intuitionistic Fuzziness and Other Intelligent Theories and Their Applications*, Studies in Computational Intelligence 757,
https://doi.org/10.1007/978-3-319-78931-6_10

1 Introduction

InterCriteria Analysis (ICrA), based on two fundamental concepts of intuitionistic fuzzy sets and index matrices gives possibility to compare some criteria or estimated by them objects and prove existing and even unexpected correlations between the criteria involved in a process of evaluation [2]. ICrA has been successfully applied in numerous areas of science and practice. Some of the recent promising ICrA applications are in the fields of economic [9], e-learning [14], algorithms performance [10, 16, 17, 20, 21], ecology [12, 13], medicine [25], etc. The idea ICrA approach to be tested in the field of fermentation processes (FP) modelling intuitively arises as a result of searching for other successful ICrA implementations. In [24] the ICrA has been implemented to establish the relations between parameters of the genetic algorithm (GA) and an *E. coli* FP model parameters. In [22] authors consider ICrA by pairs and triples for more deep criteria analysis. In [23] four different algorithms for Inter-Criteria relations calculation, namely μ-biased, Balanced, ν-biased and Unbiased, are proposed. In order to gain more generalized conclusions about the considered algorithms performance, a further investigation has been conducted in [19] but for different case study.

Meanwhile, the ICrA approach theory is continuously being developed and here a newly elaborated Weighted algorithm for criterial relations calculation is going to be proposed. In this investigation altogether five different ICrA algorithms, μ-biased, Balanced, ν-biased, Unbiased and Weighted algorithms are applied to explore the correlations between *E. coli* and *S. cerevisiae* FP model parameters, from one side, and genetic algorithms parameters, namely the number of individuals *ind* and the number of generations *gen*, from the other side. The obtained results are compared and the most reliable algorithm for ICrA relations calculation has been distinguished.

2 Problem Formulation

2.1 Case Study 1: E. coli Fed-Batch Fermentation Process

The mathematical model of the considered here *E. coli* fed-batch process is presented by the following non-linear differential equations system [18]:

$$\frac{dX}{dt} = \mu_{max} \frac{S}{k_S + S} X - \frac{F_{in}}{V} X \tag{1}$$

$$\frac{dS}{dt} = -\frac{1}{Y_{S/X}} \mu_{max} \frac{S}{k_S + S} X + \frac{F_{in}}{V} (S_{in} - S) \tag{2}$$

$$\frac{dV}{dt} = F_{in} \tag{3}$$

where

- X – biomass concentration, [g/l];
- S – substrate (glucose) concentration, [g/l];
- F_{in} – feeding rate, [l/h];
- V – bioreactor volume, [l];
- μ_{max} – maximum value of the specific growth rate, [1/h];
- k_S – saturation constant, [g/l];
- S_{in} – substrate concentration in feeding solution, [g/l];
- $Y_{S/X}$ – yield coefficient, [-].

All functions are continuous and differentiable, and all model parameters fulfil the non-zero division requirement.

The parameter vector that should be identified for the model Eqs. (1)–(3) is $p_1 = [\mu_{max}\ k_S\ Y_{S/X}]$.

Identification procedures of model parameters of an *E. coli* MC4110 fed-batch FP are performed based on experimental data for biomass and glucose concentrations. The detailed description of the processes conditions and experimental data set can be found in [18].

2.2 Case Study 2: S. cerevisiae Fed-Batch Fermentation Process

The mathematical model of *S. cerevisiae* fed-batch FP is commonly described as follows, according to the mass balance and considering mixed oxidative functional state [18]:

$$\frac{dX}{dt} = \left(\mu_{2S}\frac{S}{S+k_S} + \mu_{2E}\frac{E}{E+k_E}\right)X - \frac{F_{in}}{V}X \tag{4}$$

$$\frac{dS}{dt} = -\frac{\mu_{2S}}{Y_{S/X}}\frac{S}{S+k_S}X + \frac{F_{in}}{V}(S_{in} - S) \tag{5}$$

$$\frac{dV}{dt} = F_{in}, \tag{6}$$

where, in addition to the denotations from Eqs. (1)–(3),

- E—concentration of ethanol, [g/l];
- μ_{2S}—maximum value of the specific growth rate of substrate, [1/h];
- μ_{2E}—maximum value of the specific growth rate of ethanol, [1/h];
- k_E—saturation constant, [g/l].

By analogy to Case study 1, all functions are continuous and differentiable, and all model parameters fulfil the non-zero division requirement.

The parameter vector that should be identified for the model Eqs. (4)–(6) is $p_2 = [\mu_{2S} \; \mu_{2E} \; k_S \; k_E \; Y_{S/X}]$.

Identification procedures of model parameters of a *S. cerevisiae* fed-batch FP are performed based on experimental data for biomass, glucose and ethanol concentrations, obtained in the Institute of Technical Chemistry—University of Hannover, Germany. The detailed description of the processes conditions and experimental data set can be found in [18].

2.3 Genetic Algorithm Parameters

GAs are a stochastic global optimization technique [11]. Many operators and parameters of GA need to be tuned for a concrete optimization problem. Among the most important genetic parameters that have a significant influence on the algorithm performance, are the number of individuals (*ind*) and the number of generations (*gen*). The number of individuals determines how many chromosomes are included in the population. If there are few chromosomes, GA will have fewer opportunities to perform crossover and only a small part of the search space will be investigated. On the other hand, if there are too many chromosomes, the algorithm convergence time will logically increase. Concerning the number of generations, different authors offer different solutions depending on the solved problem. The number of generations can significantly affect the accuracy of the solution and the convergence time of the algorithm.

In this investigation the impact of *ind* and *gen* is going to be evaluated examining different values of both genetic operators (Table 1). Due to the specific peculiarities of two fed-batch FP, different strategies are applied for both *gen* and *ind* in both Case studies. The selected values of *gen* and *ind* are chosen based on the following prerequisites: (i) concerning the recommended by the literature values and trying to comprise different values in the ranges for both Case studies [11, 15]; (ii) concerning the previous authors' experience of modelling of FP using GA [17, 22, 24].

All other GA operators and parameters are set as presented in Table 1, based on the recommendations in [11, 15] and authors expertise.

2.4 Optimization Criterion

The main goal is to adjust the model parameter vectors p_1 and p_2 of the considered non-linear mathematical models, respectively, Eqs. (1)–(3) and (4)–(6), aiming at the best fit to experimental data sets. Let $Z_{\text{mod}} \stackrel{\text{def}}{=} [X_{\text{mod}} \; S_{\text{mod}}]$ are the model predictions for biomass and substrate, and $Z_{\text{exp}} \stackrel{\text{def}}{=} [X_{\text{exp}} \; S_{\text{exp}}]$ – known experimental data for biomass and substrate. Then replacing $Z = Z_{\text{mod}} - Z_{\text{exp}}$, the objective function is presented as:

Table 1 Main GA parameters and operators

Parameters and operators	Case study 1	Case study 2
Encoding	Real-coded	Binary-coded
Crossover operator	Extended intermediate recombination	Double point
Mutation operator	Real-value mutation	Bit inversion
Selection operator	Roulette well selection	
ggap	0.97	0.8
gen	50, 100, 150, 200, 250, 300	100, 200, 500, 1000
ind	50, 100, 150, 200	20, 40, 60, 80, 100
xovr	0.7	0.9
mutr	0.01	0.05

$$J = \|Z\|^2 \to \min, \tag{7}$$

where $\| \cdot \|$ denotes the ℓ^2-vector norm.

3 InterCriteria Analysis

InterCriteria analysis, based on the apparatuses of Index Matrices (IM) [3, 6–8] and Intuitionistic Fuzzy Sets (IFS) [4, 5], is given in details in [2]. Here, for completeness, the proposed idea is briefly presented.

Let the initial IM is presented in the form of Eq. (8), where, for every p, q, $(1 \leq p \leq m, 1 \leq q \leq n)$, C_p is a criterion, taking part in the evaluation; O_q—an object to be evaluated; $C_p(O_q)$—a real number (the value assigned by the pth criteria to the qth object).

$$A = \begin{array}{c|ccccc} & O_1 & \dots & O_q & \dots & O_n \\ \hline C_1 & C_1(O_1) & \dots & C_1(O_q) & \dots & C_1(O_n) \\ \vdots & \vdots & \ddots & \vdots & \ddots & \vdots \\ C_p & C_p(O_1) & \dots & C_p(O_q) & \dots & C_p(O_n) \\ \vdots & \vdots & \ddots & \vdots & \ddots & \vdots \\ C_m & C_m(O_1) & \dots & C_m(O_q) & \dots & C_m(O_n) \end{array} \tag{8}$$

Let O denotes the set of all objects being evaluated, and $C(O)$ is the set of values assigned by a given criteria C (i.e., $C = C_p$ for some fixed p) to the objects, i.e.,

$$O \overset{\text{def}}{=} \{O_1, O_2, O_3, \ldots, O_n\},$$

$$C(O) \overset{\text{def}}{=} \{C(O_1), C(O_2), C(O_3), \ldots, C(O_n)\}.$$

Let $x_i = C(O_i)$. Then the following set can be defined:

$$C^*(O) \overset{\text{def}}{=} \{\langle x_i, x_j \rangle | i \neq j \ \& \ \langle x_i, x_j \rangle \in C(O) \times C(O)\}.$$

Further, if $x = C(O_i)$ and $y = C(O_j)$, $x \prec y$ will be written iff $i < j$.

In order to find the agreement between two criteria, the vectors of all internal comparisons for each criterion are constructed, which elements fulfil one of the three relations R, $\overline{R}$ and $\tilde{R}$. The nature of the relations is chosen such that for a fixed criterion C and any ordered pair $\langle x, y \rangle \in C^*(O)$:

$$\langle x, y \rangle \in R \Leftrightarrow \langle y, x \rangle \in \overline{R} \tag{9}$$

$$\langle x, y \rangle \in \tilde{R} \Leftrightarrow \langle x, y \rangle \notin (R \cup \overline{R}) \tag{10}$$

$$R \cup \overline{R} \cup \tilde{R} = C^*(O). \tag{11}$$

For the effective calculation of the vector of internal comparisons (denoted further by $V(C)$) only the subset of $C(O) \times C(O)$ needs to be considered, namely:

$$C^{\prec}(O) \overset{\text{def}}{=} \{\langle x, y \rangle | \ x \prec y \ \& \ \langle x, y \rangle \in C(O) \times C(O),$$

due to Eqs. (9)–(11). For brevity, $c^{i,j} = \langle C(O_i), C(O_j) \rangle$.

Then for a fixed criterion C the vector of lexicographically ordered pair elements is constructed:

$$V(C) = \{c^{1,2}, c^{1,3}, \ldots, c^{1,n}, c^{2,3}, c^{2,4}, \ldots, c^{2,n}, \\ c^{3,4}, \ldots, c^{3,n}, \ldots, c^{n-1,n}\}. \tag{12}$$

In order to be more suitable for calculations, $V(C)$ is replaced by $\hat{V}(C)$, where its kth component ($1 \leq k \leq \frac{n(n-1)}{2}$) is given by:

$$\hat{V}_k(C) = \begin{cases} 1, & \text{iff } V_k(C) \in R, \\ -1, & \text{iff } V_k(C) \in \overline{R}, \\ 0, & \text{otherwise.} \end{cases}$$

When comparing two criteria C and C', the degree of "agreement" ($\mu_{C,C'}$) is usually determined as the number of matching components of the respective vectors. The degree of "disagreement" ($\nu_{C,C'}$) is usually the number of components of opposing signs in the two vectors. From the way of computation it is obvious that $\mu_{C,C'} = \mu_{C',C}$ and $\nu_{C,C'} = \nu_{C',C}$. Moreover, $\langle \mu_{C,C'}, \nu_{C,C'} \rangle$ is an Intuitionistic Fuzzy Pair (IFP).

There may be some pairs $\langle \mu_{C,C'}, \nu_{C,C'} \rangle$, for which the sum $\mu_{C,C'} + \nu_{C,C'}$ is less than 1. The difference

$$\pi_{C,C'} = 1 - \mu_{C,C'} - \nu_{C,C'} \tag{13}$$

is considered as a degree of "uncertainty".

In this investigation five different algorithms for calculation of $\mu_{C,C'}$ and $\nu_{C,C'}$ are used, based on the ideas described in [1] and here illustrated by Tables 2, 3 and 4. The pseudocode of the first four following algorithms can be found in [23].

- **μ-biased**: This algorithm follows the rules presented in Table 2, where the rule for $=, =$ for two criteria C and C' is assigned to $\mu_{C,C'}$.
- **Balanced**: This algorithm follows the rules in Table 3 [1], where the rule for $=, =$ for two criteria C and C' is assigned a half to both $\mu_{C,C'}$ and $\nu_{C,C'}$. It should be noted that in such case a criteria compared to itself does not necessarily yield $\langle 1, 0 \rangle$.
- **ν-biased**: By analogy of μ-biased algorithm in this case the rule for $=, =$ for two criteria C and C' is assigned to $\nu_{C,C'}$. It should be noted that in such case a criteria compared to itself does not necessarily yield $\langle 1, 0 \rangle$.
- **Unbiased**: This algorithm follows the rules in Table 4. It should be noted that in such case a criterion compared to itself does not necessarily yield $\langle 1, 0 \rangle$, too.
- **Weighted**: This algorithm is newly proposed and it is based on the Unbiased for the initial estimation of $\mu_{C,C'}$ and $\nu_{C,C'}$, however, at the end of it the values of $\pi_{C,C'}$ are proportionally distributed to $\mu_{C,C'}$ and $\nu_{C,C'}$. Thus, the final values of $\mu_{C,C'}$ and $\nu_{C,C'}$ generated by this algorithm will always complement to 1. An example pseudocode is presented below as Algorithm 1.

Table 2 ([1], Table 3) Calculation of μ and ν in μ-biased algorithm

$>, >$	$>, =$	$>, <$	$=, >$	$=, =$	$=, <$	$<, >$	$<, =$	$<, <$
$\mu+$	$\pi+$	$\nu+$	$\pi+$	$\mu+$	$\pi+$	$\nu+$	$\pi+$	$\mu+$

Table 3 ([1], Table 2) Calculation of μ and ν in Balanced algorithm

$>, >$	$>, =$	$>, <$	$=, >$	$=, =$	$=, <$	$<, >$	$<, =$	$<, <$
$\mu+$	$\pi+$	$\nu+$	$\pi+$	$1/2\mu+$ $1/2\nu+$	$\pi+$	$\nu+$	$\pi+$	$\mu+$

Table 4 ([1], Table 1) Calculation of μ and ν in Unbiased algorithm

$>, >$	$>, =$	$>, <$	$=, >$	$=, =$	$=, <$	$<, >$	$<, =$	$<, <$
$\mu+$	$\pi+$	$\nu+$	$\pi+$	$\pi+$	$\pi+$	$\nu+$	$\pi+$	$\mu+$

Algorithm 1 : Weighted

Require: Vectors $\hat{V}(C)$ and $\hat{V}(C')$

1: **function** DEGREES OF AGREEMENT AND DISAGREEMENT($\hat{V}(C)$, $\hat{V}(C')$)
2:     $P \leftarrow \hat{V}(C) \odot \hat{V}(C')$ ▷ $\odot$ denotes Hadamard (entrywise) product
3:     $V \leftarrow \hat{V}(C) - \hat{V}(C')$
4:     $\mu \leftarrow 0$
5:     $\nu \leftarrow 0$
6:     **for** $i \leftarrow 1$ to $\frac{n(n-1)}{2}$ **do**
7:         **if** $V_i = 0$ and $P_i \neq 0$ **then**
8:             $\mu \leftarrow \mu + 1$
9:         **else if** abs(V_i) = 2 **then**   ▷ abs(V_i): the absolute value of V_i
10:             $\nu \leftarrow \nu + 1$
11:        **end if**
12:    **end for**
13:    $\mu \leftarrow \frac{2}{n(n-1)}\mu$
14:    $\nu \leftarrow \frac{2}{n(n-1)}\nu$
15:    **if** $\mu + \nu \neq 0$ **then**
16:        $\mu \leftarrow \frac{\mu}{\mu+\nu}$
17:        $\nu \leftarrow \frac{\nu}{\mu+\nu}$
18:    **else**
19:        $\mu \leftarrow \frac{1}{2}$
20:        $\nu \leftarrow \frac{1}{2}$
21:    **end if**
22:    **return** μ, ν
23: **end function**

Table 5 Comparison of the considered here five algorithms for ICrA relations calculation

Condition	μ-biased	ν-biased	Balanced	Unbiased	Weighted
Initiate	$\mu \leftarrow 0$	$\mu \leftarrow 0$	$\mu \leftarrow 0$	$\mu \leftarrow 0$	$\mu \leftarrow 0$
	$\nu \leftarrow 0$	$\nu \leftarrow 0$	$\nu \leftarrow 0$	$\nu \leftarrow 0$	$\nu \leftarrow 0$
$P_i = V_i$	$\mu \leftarrow \mu + 1$	$\nu \leftarrow \nu + 1$	$\mu \leftarrow \mu + 1/2$		
			$\nu \leftarrow \nu + 1/2$		
$P_i \neq V_i = 0$	$\mu \leftarrow \mu + 1$	$\mu \leftarrow \mu + 1$	$\mu \leftarrow \mu + 1$	$\mu \leftarrow \mu + 1$	$\mu \leftarrow \mu + 1$
$P_i = -1$	$\nu \leftarrow \nu + 1$	$\nu \leftarrow \nu + 1$	$\nu \leftarrow \nu + 1$	$\nu \leftarrow \nu + 1$	$\nu \leftarrow \nu + 1$
Normalization	$\mu \leftarrow \dfrac{2}{n(n-1)}\mu$ $\nu \leftarrow \dfrac{2}{n(n-1)}\nu$				
Postcalculation for Weighted only					
$\mu + \nu = 0$					$\mu \leftarrow 1/2$
					$\nu \leftarrow 1/2$
$\mu + \nu \neq 0$					$\mu \leftarrow \dfrac{\mu}{(\mu + \nu)}$
					$\nu \leftarrow \dfrac{\nu}{(\mu + \nu)}$
Return	μ, ν				

For completeness, Table 5 presents a snapshot of the considered here five algorithms for ICrA relations calculation, including the newly presented one—Weighted algorithm, where P and V are defined as in Algorithm 1.

No matter which one of the considered here five algorithms is applied to IM A Eq. (8), as a result an IM of the following type is constructed:

$$
\begin{array}{c|ccc}
 & C_2 & \cdots & C_m \\
\hline
C_1 & \langle \mu_{C_1,C_2}, \nu_{C_1,C_2} \rangle & \cdots & \langle \mu_{C_1,C_m}, \nu_{C_1,C_m} \rangle \\
\vdots & \vdots & \ddots & \vdots \\
C_{m-1} & & \cdots & \langle \mu_{C_{m-1},C_m}, \nu_{C_{m-1},C_m} \rangle
\end{array}
,
$$

that determines the degrees of "agreement" and "disagreement" between criteria $C_1, ..., C_m$.

4 Numerical Results and Discussion

4.1 Case Study 1: E. coli Fed-Batch Fermentation Process

GA with different values of ind (forming four GA) and gen (forming six GA) are performed to obtain the model parameters estimates for the E. coli MC4110 fed-batch FP. 30 independent runs of each GA have been performed due to the algorithm's stochastic nature.

Based on the obtained average results from the identification procedures, as well as the average values of the GA convergence time (T) and of objective function (J), the following two IMs are constructed:

$$
A_{ind} = \begin{array}{c|cccc}
 & GA_{ind}^{1,1} & GA_{ind}^{1,2} & GA_{ind}^{1,3} & GA_{ind}^{1,4} \\
\hline
J & 0.006 & 0.003 & 0.012 & 0.004 \\
T & 109.73 & 231.88 & 562.67 & 691.48 \\
ind & 50 & 100 & 150 & 200 \\
\mu_{max} & 0.55 & 0.55 & 0.55 & 0.55 \\
k_S & 0.01 & 0.01 & 0.01 & 0.01 \\
Y_{S/X} & 2.00 & 2.00 & 2.00 & 2.00
\end{array}
\qquad (14)
$$

IM A_{ind} presents average estimates of the model parameters μ_{max}, k_S and $Y_{S/X}$, as well as the convergence time (T) and the objective function value (J), respectively in the case of 50, 100, 150 and 200 individuals, named ($GA_{ind}^{1,1} - GA_{ind}^{1,4}$). Obtained average estimates of the model parameters and convergence time are rounded up to the second digit after the decimal point, while the objective function is rounded up to the third digit.

By analogy, IM A_{gen} presents the average and rounded results for $\mu_{max}, k_S,$ $Y_{S/X}, T, J$ and gen, respectively in the case of 50, 100, 150, 200, 250 and 300 generations $(GA_{gen}^{1,1} - GA_{gen}^{1,6})$.

$$
A_{gen} = \begin{array}{c|cccccc}
 & GA_{gen}^{1,1} & GA_{gen}^{1,2} & GA_{gen}^{1,3} & GA_{gen}^{1,4} & GA_{gen}^{1,5} & GA_{gen}^{1,6} \\
\hline
J & 0.023 & 0.025 & 0.015 & 0.003 & 0.003 & 0.003 \\
T & 48.55 & 88.42 & 277.61 & 231.88 & 306.31 & 350.59 \\
gen & 50 & 100 & 150 & 200 & 250 & 300 \\
\mu_{max} & 0.54 & 0.55 & 0.55 & 0.55 & 0.55 & 0.55 \\
k_S & 0.01 & 0.01 & 0.01 & 0.01 & 0.01 & 0.01 \\
Y_{S/X} & 2.00 & 2.00 & 2.00 & 2.00 & 2.00 & 2.00
\end{array}
\tag{15}
$$

As it can be seen from the presented IMs Eqs. (14) and (15), there are many equal estimates for $C_p(O_q)$. These are good examples to test the proposed five ICrA algorithms for calculation of degrees of "agreement" ($\mu_{C,C'}$) and of "disagreement" ($\nu_{C,C'}$).

Using the mentioned above IMs Eqs. (14) and (15), the proposed five ICrA algorithms (μ-biased, Balanced, ν-biased, Unbiased and Weighted) are applied to calculate the IFP $\langle \mu_{C,C'}, \nu_{C,C'} \rangle$ for every two pairs of considered criteria. Following the Eq. (13), π-values are calculated, too. The results are summarized in the Tables 6 and 7, considering different values of ind and gen.

As it could be seen from the obtained results, the highest correlation have been observed between convergence time T and each of the investigated GA parameters

Table 6 Results from the ICrA in case of *E. coli* FP—*ind*

Criteria pairs	μ-biased			Balanced			ν-biased			Unbiased			Weighted		
	μ	ν	π	μ	ν	π	μ	ν	π	μ	ν	π	μ	ν	π
$J - T$	0.50	0.50	0	0.50	0.50	0	0.50	0.50	0	0.50	0.50	0	0.50	0.50	0
$J - ind$	0.50	0.50	0	0.50	0.50	0	0.50	0.50	0	0.50	0.50	0	0.50	0.50	0
$J - \mu_{max}$	0	0	1	0	0	1	0	0	1	0	0	1	0.50	0.50	0
$J - k_S$	0	0	1	0	0	1	0	0	1	0	0	1	0.50	0.50	0
$J - Y_{S/X}$	0	0	1	0	0	1	0	0	1	0	0	1	0.50	0.50	0
$T - ind$	1	0	0	1	0	0	1	0	0	1	0	0	1	0	0
$T - \mu_{max}$	0	0	1	0	0	1	0	0	1	0	0	1	0.50	0.50	0
$T - k_S$	0	0	1	0	0	1	0	0	1	0	0	1	0.50	0.50	0
$T - Y_{S/X}$	0	0	1	0	0	1	0	0	1	0	0	1	0.50	0.50	0
$ind - \mu_{max}$	0	0	1	0	0	1	0	0	1	0	0	1	0.50	0.50	0
$ind - k_S$	0	0	1	0	0	1	0	0	1	0	0	1	0.50	0.50	0
$ind - Y_{S/X}$	0	0	1	0	0	1	0	0	1	0	0	1	0.50	0.50	0
$\mu_{max} - k_S$	1	0	0	0.50	0.50	0	0	1	0	0	0	1	0.50	0.50	0
$\mu_{max} - Y_{S/X}$	1	0	0	0.50	0.50	0	0	1	0	0	0	1	0.50	0.50	0
$k_S - Y_{S/X}$	1	0	0	0.50	0.50	0	0	1	0	0	0	1	0.50	0.50	0

Table 7 Results from the ICrA in case of *E. coli* FP—*gen*

Criteria pairs	μ-biased			Balanced			ν-biased			Unbiased			Weighted		
	μ	ν	π	μ	ν	π	μ	ν	π	μ	ν	π	μ	ν	π
$J - T$	0.13	0.67	0.20	0.13	0.67	0.20	0.13	0.67	0.20	0.13	0.67	0.20	0.17	0.83	0
$J - gen$	0.07	0.73	0.20	0.07	0.73	0.20	0.07	0.73	0.20	0.07	0.73	0.20	0.08	0.92	0
$J - \mu_{max}$	0.27	0.27	0.47	0.17	0.37	0.47	0.07	0.47	0.47	0.07	0.27	0.67	0.20	0.80	0
$J - k_S$	0.20	0	0.80	0.10	0.10	0.80	0	0.20	0.80	0	0	1	0.50	0.50	0
$J - Y_{S/X}$	0.20	0	0.80	0.10	0.10	0.80	0	0.20	0.80	0	0	1	0.50	0.50	0
$T - gen$	0.93	0.07	0	0.93	0.07	0	0.93	0.07	0	0.93	0.07	0	0.93	0.07	0
$T - \mu_{max}$	0.33	0	0.67	0.33	0	0.67	0.33	0	0.67	0.33	0	0.67	1	0	0
$T - k_S$	0	0	1	0	0	1	0	0	1	0	0	1	0.50	0.50	0
$T - Y_{S/X}$	0	0	1	0	0	1	0	0	1	0	0	1	0.50	0.50	0
$gen - \mu_{max}$	0.33	0	0.67	0.33	0	0.67	0.33	0	1	0	0	0.67	1	0	0
$gen - k_S$	0	0	1	0	0	1	0	0	1	0	0	1	0.50	0.50	0
$gen - Y_{S/X}$	0	0	1	0	0	1	0	0	1	0	0	1	0.50	0.50	0
$\mu_{max} - k_S$	0.67	0	0.33	0.33	0.33	0.33	0	0.67	0.33	0	0	1	0.50	0.50	0
$\mu_{max} - Y_{S/X}$	0.67	0	0.33	0.33	0.33	0.33	0	0.67	0.33	0	0	1	0.50	0.50	0
$k_S - Y_{S/X}$	1	0	0	0.50	0.50	0	0	1	0	0	0	1	0.50	0.50	0

ind and *gen* ($T - gen$), as well as between pairs of model parameters themselves ($\mu_{max} - k_S$, $\mu_{max} - Y_{S/X}$ and $k_S - Y_{S/X}$). The strongest correlations between T and GA parameters ($\mu = 1$ in the case of *ind* and $\mu = 0.93$ in the case of *gen*, in the sense of strong positive and positive consonance [1], respectively) have been expected due to the fact that the convergence time logically increases when the bigger number of individuals in the population is chosen or greater number of generations is set to find a solution. The other observed strong dependencies, namely for the criteria pairs $\mu_{max} - k_S$, $\mu_{max} - Y_{S/X}$ and $k_S - Y_{S/X}$, have been also expected, since they are caused by the physical meaning of FP models parameters.

Looking for other coincidences for both GA parameters *ind* and *gen*, it should be noted that there is low or even lack of correlations between GA outcomes J, T and GA parameters *ind*/*gen*, on the one hand, and FP model parameters, on the other. For the rest of the criteria pairs $J - T$ and $J - ind/gen$, there are no clearly discernible tendencies of correlations, but it can be stated that GA parameter *ind* renders greater influence to the value of the objective function J than GA parameter *gen*.

It is interesting to note that the first four different algorithms implemented in ICrA uniquely define μ-values for 12 out of 15 criteria pairs, when the influence of GA parameter *ind* has been investigated. For the rest three criteria pairs, namely $\mu_{max} - k_S$, $\mu_{max} - Y_{S/X}$ and $k_S - Y_{S/X}$, the value of μ varies depending on the applied ICrA algorithm. This fact might be explained by the equal evaluations of FP model parameters, as it can be seen from the input IMs, and, respectively, with the different assignment of the rule for $=$, $=$ by the five different algorithms.

It is worth noting that all algorithms give identical evaluations for degrees of "agreement" and degrees of "disagreement" for four criteria pair, namely $J - T$, $J - ind$ and $T - ind$ with respect to parameter ind, and $T - gen$ with respect to parameter gen.

When the influence of GA parameter gen has been investigated, the number of criteria pairs with equal μ-values obtained after application of the first four algorithms decreases to 9, while the number of those with different values increases to 6. This is caused by another equal estimates of J in the input IMs.

It is obvious that the different proposed algorithms for calculation of the degrees of "agreement" ($\mu_{C,C'}$) and degrees of "disagreement" ($\nu_{C,C'}$) there are discrepancies only for the those criteria that have equal estimates by pairs. Considering ind (Table 6), ICrA results show different degrees of "agreement" and "disagreement" only for criteria pairs $\mu_{max} - k_S$, $\mu_{max} - Y_{S/X}$ and $k_S - Y_{S/X}$. The most reliable results, from the authors expertise, are those obtained by μ-biased algorithm. For the mentioned above criteria pairs, Balanced algorithm shows lower μ- values, i.e. weaker criteria pair relations, which contradicts to the physical meaning of the considered criteria. For the same criteria pairs (except for $T - \mu_{max}$), ν-biased algorithm and Unbiased algorithm calculate $\mu = 0$ (in the sense of strong negative consonance [1]), which are the opposite results. Moreover, Unbiased algorithm gives the π-value equal to 1 (except $T - \mu_{max}$), which is the maximum degree of "uncertainty".

Similar results are obtained with respect to the parameter gen (Table 7). In this case ICrA results show discrepancies again for criteria pairs $\mu_{max} - k_S$, $\mu_{max} - Y_{S/X}$ and $k_S - Y_{S/X}$, as well as for the criteria pairs $T - \mu_{max}$, $T - k_S$ and $J - Y_{S/X}$, caused by the equal values of the objective function J.

As it has been mentioned above, the higher degrees of "uncertainty" π are caused by the equivalent estimations of the FP model parameters, which definitely is not a negative result. In opposite, this is an evidence of the workability and stability of the genetic algorithms as a reliable optimization technique. The newly elaborated and presented here Weighted algorithm aims namely to eliminate the degree of "uncertainty" π. Looking at the results obtained by the application of Weighted algorithm, it is interesting to be noted that there are altogether 5 cases with the highest value of degree of "agreement" $\mu = 1$ – four in the investigation of ind, namely for the pairs of $T - ind$, $\mu_{max} - k_S$, $\mu_{max} - Y_{S/X}$ and $k_S - Y_{S/X}$, and 1 in gen, namely $k_S - Y_{S/X}$. Due to the fact that the Weighted algorithm is based on the Unbiased one, in 4 out of 5 mentioned above pairs, when the Unbiased algorithm gives the degree of "uncertainty" $\pi = 1$, the Weighted algorithm distributes the value of $\mu = 1$ equally to $\mu = 0.5$ and $\nu = 0.5$, while, if the Unbiased algorithm keeps the $\mu = 1$, the Weighted algorithm also keeps $\mu = 1$ (case of $T - ind$ in the investigation of ind). There are also a lot of pairs—altogether 9 pairs in the investigation of ind and 4 in the investigation of gen, in which the degree of "uncertainty" $\pi = 1$ is overcame and the Weighted algorithm distributes the value of $\pi = 1$ equally to $\mu = 0.5$ and $\nu = 0.5$. So as it is evident, the application of Weighted algorithm is "double edged sword"—one can eliminate degree of "uncertainty" $\pi = 1$ gaining $\mu = 0.5$ and $\nu = 0.5$, or loosing of degree of "agreement" $\mu = 1$, again gaining $\mu = 0.5$ and $\nu = 0.5$.

As a summary of the thorough analysis presented so far, it may be summarised that the μ-biased algorithm is the most reliable one.

4.2 Case Study 2: S. cerevisiae Fed-Batch Fermentation Process

GA with different values of ind (forming five GA) and gen (forming four GA) are consequently applied to a model parameter identification of S. cerevisiae fed-batch FP. Due to the algorithm's stochastic nature 30 independent runs of each GA have been performed. Based on the obtained average results for the model parameters, GA convergence time (T) and objective function (J), the two IMs A_{ind} and A_{gen} are constructed:

$$A_{ind} = \begin{array}{c|ccccc} & GA_{ind}^{1,1} & GA_{ind}^{1,2} & GA_{ind}^{1,3} & GA_{ind}^{1,4} & GA_{ind}^{1,5} \\ \hline J & 0.022 & 0.022 & 0.022 & 0.022 & 0.022 \\ T & 79.2 & 145.5 & 212.8 & 295.5 & 375.1 \\ ind & 20 & 40 & 60 & 80 & 100 \\ \mu_{2S} & 0.95 & 0.97 & 0.95 & 0.97 & 0.98 \\ \mu_{2E} & 0.12 & 0.12 & 0.12 & 0.14 & 0.14 \\ k_S & 0.13 & 0.13 & 0.12 & 0.13 & 0.13 \\ k_E & 0.8 & 0.8 & 0.8 & 0.8 & 0.8 \\ Y_{S/X} & 0.41 & 0.41 & 0.41 & 0.4 & 0.4 \end{array} \qquad (16)$$

$$A_{gen} = \begin{array}{c|cccc} & GA_{gen}^{1,1} & GA_{gen}^{1,2} & GA_{gen}^{1,3} & GA_{gen}^{1,4} \\ \hline J & 0.022 & 0.022 & 0.022 & 0.022 \\ T & 76.8 & 148.1 & 368.3 & 771.3 \\ gen & 100 & 200 & 500 & 1000 \\ \mu_{2S} & 0.96 & 0.95 & 0.95 & 0.97 \\ \mu_{2E} & 0.13 & 0.12 & 0.12 & 0.14 \\ k_S & 0.13 & 0.12 & 0.12 & 0.13 \\ k_E & 0.8 & 0.8 & 0.8 & 0.8 \\ Y_{S/X} & 0.41 & 0.42 & 0.41 & 0.4 \end{array} \qquad (17)$$

IM A_{ind} consists of average estimations of the five model parameters μ_{2S}, μ_{2E}, k_S, k_E, $Y_{S/X}$, as well as of T and J, respectively in the case of $ind = \{20, 40, 60, 80, 100\}$, corresponding to $GA_{ind}^{1,1} \div GA_{ind}^{1,5}$. Average estimations of the model parameters and objective function values are rounded up to the second digit after the decimal point, while the convergence time is rounded up to the first digit. IM A_{gen} is constructed in a similar manner. It consists of the average and rounded results for μ_{2S}, μ_{2E}, k_S, k_E, $Y_{S/X}$, T, and J, respectively in the case of $gen = \{100, 200, 500, 1000\}$, corresponding to $GA_{gen}^{1,1} \div GA_{gen}^{1,4}$.

In the presented IMs Eqs. (16) and (17), there are many equal estimates for J and model parameters. As such, they are good examples to test the proposed five algorithms for InterCriteria relations calculation of $\mu_{C,C'}$, $\nu_{C,C'}$ and $\pi_{C,C'}$.

Five algorithms for InterCriteria relations calculation (μ-biased, Balanced, ν-biased, Unbiased and Weighted) have been applied to constructed above two IMs Eqs. (16) and (17). Thus, the IF pairs $\langle \mu_{C,C'}, \nu_{C,C'} \rangle$ and $\pi_{C,C'}$ for every two pairs of considered criteria are calculated. The results are listed in Tables 8 and 9, respectively for different values of ind and gen.

Table 8 Results from the ICrA in case of *S. cerevisiae* FP – ind

Criteria pairs	μ-biased			Balanced			ν-biased			Unbiased			Weighted		
	μ	ν	π	μ	ν	π	μ	ν	π	μ	ν	π	μ	ν	π
$J-T$	0	0	1	0	0	1	0	0	1	0	0	1	0.50	0.50	0
$J-ind$	0	0	1	0	0	1	0	0	1	0	0	1	0.50	0.50	0
$J-\mu_{2S}$	0.17	0	0.83	0.08	0.08	0.83	0	0.17	0.83	0	0	1	0.50	0.50	0
$J-\mu_{2E}$	0.17	0	0.83	0.08	0.08	0.83	0	0.17	0.83	0	0	1	0.50	0.50	0
$J-k_S$	0.33	0	0.67	0.17	0.17	0.67	0	0.33	0.67	0	0	1	0.50	0.50	0
$J-k_E$	1	0	0	0.50	0.50	0	0	1	0	0	0	1	0.50	0.50	0
$J-Y_{S/X}$	0.17	0	0.83	0.08	0.08	0.83	0	0.17	0.83	0	0	1	0.50	0.50	0
$T-ind$	1	0	0	1	0	0	1	0	0	1	0	0	1	0	0
$T-\mu_{2S}$	0.50	0.33	0.17	0.50	0.33	0.17	0.50	0.33	0.17	0.50	0.33	0.17	0.88	0.12	0
$T-\mu_{2E}$	0.50	0.33	0.17	0.50	0.33	0.17	0.50	0.33	0.17	0.50	0.33	0.17	1	0	0
$T-k_S$	0.33	0.33	0.33	0.33	0.33	0.33	0.33	0.33	0.33	0.33	0.33	0.33	0.50	0.50	0
$T-k_E$	0	0	1	0	0	1	0	0	1	0	0	1	0.50	0.50	0
$T-Y_{S/X}$	0.17	0.67	0.17	0.17	0.67	0.17	0.17	0.67	0.17	0.17	0.67	0.17	0	1	0
$ind-\mu_{2S}$	0.50	0.33	0.17	0.50	0.33	0.17	0.50	0.33	0.17	0.50	0.33	0.17	0.88	0.12	0
$ind-\mu_{2E}$	0.50	0.33	0.17	0.50	0.33	0.17	0.50	0.33	0.17	0.50	0.33	0.17	1	0	0
$ind-k_S$	0.33	0.33	0.33	0.33	0.33	0.33	0.33	0.33	0.33	0.33	0.33	0.33	0.50	0.50	0
$ind-k_E$	0	0	1	0	0	1	0	0	1	0	0	1	0.50	0.50	0
$ind-Y_{S/X}$	0.17	0.67	0.17	0.17	0.67	0.17	0.17	0.67	0.17	0.17	0.67	0.17	0	1	0
$\mu_{2S}-\mu_{2E}$	1	0	0	0.92	0.08	0	0.83	0.17	0	0.83	0	0.17	1	0	0
$\mu_{2S}-k_S$	0.83	0	0.17	0.75	0.08	0.17	0.67	0.17	0.17	0.67	0	0.33	1	0	0
$\mu_{2S}-k_E$	0.17	0	0.83	0.08	0.08	0.83	0	0.17	0.83	0	0	1	0.50	0.50	0
$\mu_{2S}-Y_{S/X}$	0	0.67	0.33	0	0.67	0.33	0	0.67	0.33	0	0.67	0.33	0	1	0
$\mu_{2E}-k_S$	0.83	0	0.17	0.75	0.08	0.17	0.67	0.17	0.17	0.67	0	0.33	1	0	0
$\mu_{2E}-k_E$	0.17	0	0.83	0.08	0.08	0.83	0	0.17	0.83	0	0	1	0.50	0.50	0
$\mu_{2E}-Y_{S/X}$	0	0.67	0.33	0	0.67	0.33	0	0.67	0.33	0	0.67	0.33	0	1	0
k_S-k_E	0.33	0	0.67	0.17	0.17	0.67	0	0.33	0.67	0	0	1	0.50	0.50	0
$k_S-Y_{S/X}$	0	0.50	0.50	0	0.50	0.50	0	0.50	0.50	0	0.50	0.50	0	1	0
$k_E-Y_{S/X}$	0.17	0	0.83	0.08	0.08	0.83	0	0.17	0.83	0	0	1	0.50	0.50	0

Table 9 Results from the ICrA in case of *S. cerevisiae* FP—*gen*

Criteria pairs	μ-biased			Balanced			ν-biased			Unbiased			Weighted		
	μ	ν	π	μ	ν	π	μ	ν	π	μ	ν	π	μ	ν	π
$J - T$	0	0	1	0	0	1	0	0	1	0	0	1	0.50	0.50	0
$J - gen$	0	0	1	0	0	1	0	0	1	0	0	1	0.50	0.50	0
$J - \mu_{2S}$	0.20	0	0.80	0.10	0.10	0.80	0	0.20	0.80	0	0	1	0.50	0.50	0
$J - \mu_{2E}$	0.40	0	0.60	0.20	0.20	0.60	0	0.40	0.60	0	0	1	0.50	0.50	0
$J - k_S$	0.60	0	0.40	0.30	0.30	0.40	0	0.60	0.40	0	0	1	0.50	0.50	0
$J - k_E$	1	0	0	0.50	0.50	0	0	1	0	0	0	1	0.50	0.50	0
$J - Y_{S/X}$	0.40	0	0.60	0.20	0.20	0.60	0	0.40	0.60	0	0	1	0.50	0.50	0
$T - gen$	1	0	0	1	0	0	1	0	0	1	0	0	1	0	0
$T - \mu_{2S}$	0.70	0.10	0.20	0.70	0.10	0.20	0.70	0.10	0.20	0.70	0.10	0.20	0.60	0.40	0
$T - \mu_{2E}$	0.60	0	0.40	0.60	0	0.40	0.60	0	0.40	0.60	0	0.40	0.60	0.40	0
$T - k_S$	0.20	0.20	0.60	0.20	0.20	0.60	0.20	0.20	0.60	0.20	0.20	0.60	0.50	0.50	0
$T - k_E$	0	0	1	0	0	1	0	0	1	0	0	1	0.50	0.50	0
$T - Y_{S/X}$	0	0.60	0.40	0	0.60	0.40	0	0.60	0.40	0	0.60	0.40	0.20	0.80	0
$gen - \mu_{2S}$	0.70	0.10	0.20	0.70	0.10	0.20	0.70	0.10	0.20	0.70	0.10	0.20	0.60	0.40	0
$gen - \mu_{2E}$	0.60	0	0.40	0.60	0	0.40	0.60	0	0.40	0.60	0	0.40	0.60	0.40	0
$gen - k_S$	0.20	0.20	0.60	0.20	0.20	0.60	0.20	0.20	0.60	0.20	0.20	0.60	0.50	0.50	0
$gen - k_E$	0	0	1	0	0	1	0	0	1	0	0	1	0.50	0.50	0
$gen - Y_{S/X}$	0	0.60	0.40	0	0.60	0.40	0	0.60	0.40	0	0.60	0.40	0.20	0.80	0
$\mu_{2S} - \mu_{2E}$	0.60	0	0.40	0.55	0.05	0.40	0.50	0.10	0.40	0.50	0	0.50	1	0	0
$\mu_{2S} - k_S$	0.40	0	0.60	0.35	0.05	0.60	0.30	0.10	0.60	0.30	0	0.70	1	0	0
$\mu_{2S} - k_E$	0.20	0	0.80	0.10	0.10	0.80	0	0.20	0.80	0	0	1	0.50	0.50	0
$\mu_{2S} - Y_{S/X}$	0.10	0.50	0.40	0.05	0.55	0.40	0	0.60	0.40	0	0.50	0.50	0	1	0
$\mu_{2E} - k_S$	0.40	0	0.60	0.30	0.10	0.60	0.20	0.20	0.60	0.20	0	0.80	1	0	0
$\mu_{2E} - k_E$	0.40	0	0.60	0.20	0.20	0.60	0	0.40	0.60	0	0	1	0.50	0.50	0
$\mu_{2E} - Y_{S/X}$	0.40	0.60	0	0.20	0.80	0	0	1	0	0	0.60	0.40	0	1	0
$k_S - k_E$	0.60	0	0.40	0.30	0.30	0.40	0	0.60	0.40	0.5	0.5	0	0.50	0.50	0
$k_S - Y_{S/X}$	0.20	0.20	0.60	0.10	0.30	0.60	0	0.40	0.60	0	0.20	0.80	0	1	0
$k_E - Y_{S/X}$	0.40	0	0.60	0.20	0.20	0.60	0	0.40	0.60	0	0	1	0.50	0.50	0

The results from all five investigated algorithms show, that the strongest correlation ($\mu = 1$) (i.e. positive consonance) has been observed between convergence time $T - ind$ and $T - gen$. Such correlation had been expected, since the convergence time logically increases with the increase of number of individuals or number of generations. The results are similar to those obtained in Case study 1, although GA parameter ind renders slightly greater influence to T than to gen.

Very strong correlation ($\mu = 0$), i.e. negative consonance, has been also observed for $J - ind$ and $J - gen$, respectively. These results have been also expected since the objective function value logically decreases when the number of individuals or the number of generations increase. In other words, a solution with high accuracy is achieved either with investigation of a bigger part of the search space, or with performance of more algorithm generations, respectively. In comparison to the results obtained for Case study 1, such correlation is observed only for GA parameter gen. It is worth noting that in this case the calculated values of the degrees of "uncertainty" are $\pi_{C,C'} = 1$ due to the obtained equal values of J during all performed parameter identification procedures. From the other point of view, this is the evidence of excellent GA performance reaching the lowest J-value ($J = 0.022$) for all considered GA parameters ind and gen.

In case of μ-biased algorithm another strong correlation has been observed between J and the model parameter k_E for both investigated GA parameters ind and gen. Such a result shows a high sensitivity of k_E in comparison to the other model parameters in the vector $p_2 = [\mu_{2S}\ \mu_{2E}\ k_S\ k_E\ Y_{S/X}]$ of the non-linear mathematical model Eqs. (4)–(6).

In the case of ind variation a strong correlation between μ_{2S} and μ_{2E} is observed. μ-biased algorithm assigns the highest value to the $\mu = 1$, i.e. strong positive consonance, Balanced algorithm—the value of $\mu = 0.92$, i.e. positive consonance, while ν-biased algorithm and Unbiased algorithm—the value of $\mu = 0.83$, i.e. weak positive consonance. The pairs $\mu_{2S} - k_S$ and $\mu_{2E} - k_S$ show also $\mu = 0.83$, i.e. weak positive consonance. These high μ-values are also expected due to the physical meaning of FP models parameters [18]. For the case of GA parameter gen, the obtained results are not so convinced.

The results show that the first four algorithms for InterCriteria relations calculation uniquely define μ-values for 13 out of 28 criteria pairs, when the influence of both GA parameter ind and gen has been investigated. For the rest 15 criteria pairs, the value of μ varies depending on the applied ICrA algorithms, due to different assignment of the rule for $=, =$. Investigated in this paper different ICrA algorithms show discrepancies only for those criteria that have equal estimates by pairs, concerning input IMs Eqs. (16) and (17). As can be seen from Tables 8 and 9, different degrees of "agreement" and "disagreement" are obtained between J, T and model parameters $\mu_{2S}, \mu_{2E}, k_S, k_E$ and $Y_{S/X}$, as well as between model parameters themselves.

The application of the newly elaborated Weighted algorithm leads to results similar to those obtained in Case study 1. In Case study 2 there are altogether 5 cases with the highest value of degree of "agreement" $\mu = 1$—three in the investigation of ind, namely for the pairs of $J - k_E$, $T - ind$ and $\mu_{2S} - \mu_{2E}$, and 2 in gen, namely $J - k_E$ and $T - gen$. Again due to the fact that the Weighted algorithm is based on the Unbiased one, in 2 out of 5 mentioned above pairs, when the Unbiased algorithm gives the degree of "uncertainty" $\pi = 1$, the Weighted algorithm distributes the value of $\mu = 1$ equally to $\mu = 0.5$ and $\nu = 0.5$, while, if the Unbiased algorithm keeps the $\mu = 1$ or close to 1, the Weighted algorithm also keeps $\mu = 1$ (cases of $T - ind$ and $\mu_{2S} - \mu_{2E}$ in the investigation of ind and $T - gen$ in the investigation of gen). There are also a lot of pairs—altogether 4 pairs in the investigation of ind and 4 in the investigation of gen, in which the degree of "uncertainty" $\pi = 1$ is overcame and the Weighted algorithm distributes the value of $\pi = 1$ equally to $\mu = 0.5$ and $\nu = 0.5$—results similar to Case study 1.

As per authors expertise, comparing all considered here five algorithms for Inter-Criteria relations calculation, the most reliable results which do not contradict to the model parameters physical meaning Eqs. (1)–(3 and (4)–(6), are those obtained by μ-biased algorithm.

5 Conclusion

In this paper the InterCriteria Analysis approach is applied to establish relations and dependencies between two of the main GA parameters ind and gen, as well as the convergence time, accuracy and parameters in the considered models. The investigations have been performed for two Case studies, namely for the purposes of a parameter identification of a fed-batch FP of bacteria *E. coli* and a fed-batch FP of yeast *S. cerevisiae*. Several GA are applied with different values for ind and gen.

Five algorithms for criterial relation calculations in ICrA, namely μ-biased, Balanced, ν-biased, Unbiased and newly elaborated Weighted algorithm, are applied to parameter identification procedures. The mentioned above five algorithms for calculation of degrees of "agreement" ($\mu_{C,C'}$) and "disagreement" ($\nu_{C,C'}$) of the criteria pairs are compared based on real experimental data sets of both considered fed-batch FP processes. From the point of view of the GA performance, strong correlations have been observed between $T - ind$ and $T - gen$, as well as between $J - ind$ and $J - gen$, which is logically expected due to the GA nature.

All obtained results show that for the considered here Case studies of *E. coli* and *S. cerevisiae* fed-batch FP parameter identification, the most reliable is the μ-biased algorithm, thus making it the favourable one for the purposes of the analysis of parameter identification results considering fermentation processes. In future work, new algorithms for criterial relation calculations in ICrA might be elaborated and examined for different applications.

Acknowledgements This work was supported by the Bulgarian National Scientific Fund under the grants DFNI-I-02-5 "InterCriteria Analysis—A New Approach to Decision Making" and DN-02/10 "New Instruments for Knowledge Discovery from Data, and their Modelling."

References

1. Atanassov, K., Atanassova, V., Gluhchev, G.: InterCriteria analysis: ideas and problems. Notes on Intuitionistic Fuzzy Sets 21(1), pp. 81–88, (2015).
2. Atanassov, K., Mavrov, D., Atanassova, V.: Intercriteria decision making: a new approach for multicriteria decision making, based on index matrices and intuitionistic fuzzy sets. Issues in IFSs and GNs 11, pp. 1–8, (2014).
3. Atanassov, K.: Index matrices: Towards an augmented matrix calculus, Studies in Computational Intelligence, (2014).

4. Atanassov, K.: Intuitionistic Fuzzy Sets, VII ITKR Session, Sofia, 20-23 June 1983 (Deposed in Centr. Sci.-Techn. Library of the Bulg. Acad. of Sci., 1697/84) (in Bulgarian). Reprinted: Int J Bioauto, 2016, 20(S1), S1-S6, (2016).
5. Atanassov, K.: On Intuitionistic Fuzzy Sets Theory. Springer, Berlin, (2012).
6. Atanassov, K.: Generalized index matrices. Comptes rendus de l'Academie Bulgare des Sciences 40(11), pp. 15–18, (1987).
7. Atanassov, K.: On index matrices, part 1: standard cases. Advanced Studies in Contemporary Mathematics 20(2), pp. 291–302, (2010).
8. Atanassov, K.: On index matrices, part 2: intuitionistic fuzzy case. Proceedings of the Jangjeon Mathematical Society 13(2), pp. 121–126, (2010).
9. Atanassova, V. , Doukovska, L., Atanassov, K., Mavrov, D.: Intercriteria decision making approach to EU member states competitiveness analysis. In: Proc. of the International Symposium on Business Modeling and Software Design - BMSD'14, B. Shishkov, Ed., pp. 289–294, (2014).
10. Fidanova, S., Roeva, O., Mucherino, A., Kapanova, K.: InterCriteria analysis of ant algorithm with environment change for GPS surveying problem. Lecture Notes on Computer Science 9883, pp. 271–278, (2016).
11. Goldberg, D. E.: Genetic Algorithms in Search, Optimization and Machine Learning. Addison Wesley Longman, London, (2006).
12. Ilkova, T., Petrov, M.: Application of intercriteria analysis to the Mesta river pollution modelling. Notes on Intuitionistic Fuzzy Sets 21(2), pp. 118–125, (2015).
13. Ilkova, T., Petrov, M.: InterCriteria analysis for evaluation of the pollution of the Struma river in the Bulgarian section. Notes on IFSs 22(3), pp. 120–130, (2016).
14. Krawczak, M., Bureva, V., Sotirova, E., Szmidt, E.: Application of the InterCriteria decision making method to universities ranking. Advances in Intelligent Systems and Computing 401, pp. 365–372, (2016).
15. Obitko, M.: Genetic Algorithms. Available at http://www.obitko.com/tutorials/genetic-algorithms/, (2005).
16. Pencheva T., Angelova, M.: InterCriteria analysis of simple genetic algorithms performance. Studies in Computational Intelligence 681, pp. 147–159, (2017).
17. Pencheva, T., Angelova, M., Vassilev, P., Roeva, O.: InterCriteria analysis approach to parameter identification of a fermentation process model. Advances in Intelligent Systems and Computing 401, pp. 385–397, (2016).
18. Pencheva, T., Roeva, O., Hristozov, I.: Functional State Approach to Fermentation Processes Modelling. Prof. Marin Drinov Academic Publishing House, Sofia, (2006).
19. Pencheva, T., Roeva, O., Angelova, M.: Investigation of genetic algorithm performance based on different algorithms for intercriteria relations calculation. Lecture Notes in Computer Science 10665, pp. 390–398, (2018).
20. Roeva, O., Vassilev, P.: InterCriteria analysis of generation gap influence on genetic algorithms performance. Advances in Intelligent Systems and Computing 401, pp. 301–313, (2016).
21. Roeva, O., Fidanova, S., Paprzycki, M.: InterCriteria analysis of ACO and GA hybrid algorithms. Studies in Computational Intelligence 610, pp. 107–126, (2016).
22. Roeva, O., Pencheva, T., Angelova, M., Vassilev, P.: InterCriteria analysis by pairs and triples of genetic algorithms application for models identification. Studies in Computational Intelligence 655, pp. 193–218, (2016).
23. Roeva, O., Vassilev, P., Angelova, M., Su, J., Pencheva, T.: Comparison of different algorithms for InterCriteria relations calculation. Proc. of the 8th International Conference on Intelligent Systems, pp. 567–572, (2016)
24. Roeva, O., Vassilev, P., Fidanova, S., Paprzycki, M.: InterCriteria analysis of genetic algorithms performance. Studies in Computational Intelligence 655, pp. 235–260, (2016).
25. Todinova, S., Mavrov, D., Krumova, S., Marinov, P., Atanassova, V., Atanassov, K., Taneva, S. G.: Blood plasma thermograms dataset analysis by means of intercriteria and correlation analyses for the case of colorectal cancer. Int J Bioautomation 20(1), pp. 115–124, (2016).

Defining Consonance Thresholds in InterCriteria Analysis: An Overview

Lyubka Doukovska, Vassia Atanassova, Evdokia Sotirova, Ivelina Vardeva and Irina Radeva

Abstract The present paper aims to provide an overview of the development of the approaches adopted in defining the consonance thresholds in the recently proposed method for decision support named InterCriteria Analysis (ICA). Discussing the rationale of this leg of the ICA research, and the motivation behind each of the subsequent steps, we trace here the gradual progress in defining the thresholds of the membership and non-membership parts of the intuitionistic fuzzy pairs serving as estimations of the pairwise consonances. This progress is based on both our deepening understanding of the ICA method, and the constant observations being made during the application of ICA to a wide range of different real-life problems and datasets.

Keywords Intercriteria analysis · Decision making
Multicriteria decision making · Intuitionistic fuzzy sets · Uncertainty · Thresholds

1 Introduction

InterCriteria Analysis (ICA) is a novel mathematical method, based on the paradigms of intuitionistic fuzzy sets and index matrices, which has been recently developed in Bulgaria with the aim to support decision making in multiobject multicriteria problems. In the originally formulated problem that gave rise to the method, a part of

L. Doukovska · I. Radeva
Intelligent Systems Department, Institute of Information and Communication Technologies, Bulgarian Academy of Sciences, 2 Acad. G. Bonchev Str., Sofia 1113, Bulgaria

V. Atanassova (✉)
Bioinformatics and Mathematical Modelling Department, Institute of Biophysics and Biomedical Engineering, Bulgarian Academy of Sciences, 105 Acad. G. Bonchev Str., Sofia 1113, Bulgaria
e-mail: vassia.atanassova@gmail.com

E. Sotirova · I. Vardeva
Intelligent Systems Laboratory, "Prof. Asen Zlatarov" University, 1 "Prof. Yakimov" Blvd., Burgas 8010, Bulgaria

© Springer International Publishing AG, part of Springer Nature 2019 161
M Hadjiski and K T Atanassov (eds.), *Intuitionistic Fuzziness and Other Intelligent Theories and Their Applications*, Studies in Computational Intelligence 757,
https://doi.org/10.1007/978-3-319-78931-6_11

the criteria in an industrial multicriteria decision making problem exhibit high complexity and cost of the measurement compared to the other criteria. The assignment is to develop a method for identification of strong enough correlations between the cost-unfavourable criteria and the cost-favourable ones, in order to justifiably skip measurements against these cost-unfavourable criteria for at least part of the objects, and thus make the whole decision making process faster or cheaper. For the sake of terminological precision, in ICA, the term "correlation" is being replaced to the terms "positive/negative consonance" or "dissonance," inspired from the field of cognitive maps.

As input data, the method requires an $m \times n$ table with the measurements or evaluations of m objects against n criteria. As a result, it returns an $n \times n$ table with intuitionistic fuzzy pairs, defining the degrees of consonance between each pair of criteria, hence the name "intercriteria." Alternatively, for easier manipulation, the developed ICA software returns two $n \times n$ tables with the membership and the non-membership parts of the respective intuitionistic fuzzy pairs.

The algorithm is completely data driven and dependent on the input data from the measurements, and its present so far works well with complete datasets, without any missing values. The essence of the method is in the exhaustive pairwise comparison of the values of the measurements of all objects in the set against pairs of criteria, with all possible pairs being traversed, while counters being maintained for the percentage of the cases when the relations between the pairs of evaluations have been 'greater than,' 'less than' or 'equal.'

The method has been proposed and described in details in 2014 [1], and extensively researched in the next two years in theoretical aspect (e.g. [2–6]), with a software application being developed (see [7, 8]). The ICA method has been extensively researched not only in the light of the originally formulated industrial problem (see [9, 10]), but also for its applicability to various multicriteria multiobjects problems (e.g. [11–14]) and with the aim of improving the performance of different procedures for mathematical optimization (e.g. [15–19]). However, ICA is still being a very new field of research, giving opportunities for discussion, comparison, approbation, validation and testing with different datasets.

2 Basic Concepts and Method

The ICA method is based on two fundamental concepts: intuitionistic fuzzy sets and index matrices. Intuitionistic fuzzy sets defined by Atanassov (see [11, 20–22]) represent an extension of the concept of fuzzy sets, as defined by Zadeh [23], exhibiting function $\mu_A(x)$ defining the membership of an element x to the set A, evaluated in the [0; 1]-interval. The difference between fuzzy sets and intuitionistic fuzzy sets (IFSs) is in the presence of a second function $\nu_A(x)$ defining the non-membership of the element x to the set A, where $\mu_A(x)$ [0; 1], $\nu_A(x)$ [0; 1], and moreover $(\mu_A(x) + \nu_A(x))$ [0; 1].

The second concept on which the proposed method relies is the concept of index matrix, a matrix which features two index sets. The theory behind the index matrices is originally described in [24] and elaborated in details in [25]. Here we will start with the index matrix M with index sets with m rows $\{C_1, ..., C_m\}$ and n columns $\{O_1, ..., O_n\}$:

$$
M = \begin{array}{c|ccccccc}
 & O_1 & \cdots & O_k & \cdots & O_l & \cdots & O_n \\
\hline
C_1 & a_{C_1,O_1} & \cdots & a_{C_1,O_k} & \cdots & a_{C_1,O_l} & \cdots & a_{C_1,O_n} \\
\vdots & \vdots & \ddots & \vdots & \ddots & \vdots & \ddots & \vdots \\
C_i & a_{C_i,O_1} & \cdots & a_{C_i,O_k} & \cdots & a_{C_i,O_l} & \cdots & a_{C_i,O_n} \\
\vdots & \vdots & \ddots & \vdots & \ddots & \vdots & \ddots & \vdots \\
C_j & a_{C_j,O_1} & \cdots & a_{C_j,O_k} & \cdots & a_{C_j,O_l} & \cdots & a_{C_j,O_n} \\
\vdots & \vdots & \ddots & \vdots & \ddots & \vdots & \ddots & \vdots \\
C_m & a_{C_m,O_1} & \cdots & a_{C_m,O_j} & \cdots & a_{C_m,O_l} & \cdots & a_{C_m,O_n}
\end{array} \quad ,
$$

where for every p, q $(1 \leq p \leq m, 1 \leq q \leq n)$, C_p is a criterion (in our case, one of the twelve pillars), O_q in an evaluated object (in our case, one of the EU28 member states), $a_{C_p O_q}$ is the evaluation of the q-th object against the p-th criterion, and it is defined as a real number or another object that is comparable according to relation R with all the rest elements of the index matrix M, so that for each i, j, k it holds the relation $R\left(a_{C_k O_i}, a_{C_k O_j}\right)$. The relation R has dual relation $\bar{R}$, which is true in the cases when relation R is false, and vice versa.

For the needs of our method, pairwise comparisons between every two different criteria are made along all evaluated objects. During the comparison, it is maintained one counter of the number of times when the relation R holds, and another counter for the dual relation. Let $S_{k,l}^{\mu}$ be the number of cases where the relations $R\left(a_{C_k O_i}, a_{C_k O_j}\right)$ and $R\left(a_{C_l O_i}, a_{C_l O_j}\right)$ are simultaneously satisfied. Let also $S_{k,l}^{\nu}$ be the number of cases in which the relations $R\left(a_{C_k O_i}, a_{C_k O_j}\right)$ and its dual $\bar{R}\left(a_{C_l O_i}, a_{C_l O_j}\right)$ are simultaneously satisfied. As the total number of pairwise comparisons between the object is $n(n-1)/2$, it is seen that there hold the inequalities:

$$
0 \leq S_{k,l}^{\mu} + S_{k,l}^{\nu} \leq \frac{n(n-1)}{2}.
$$

For every k, l, such that $1 \leq k \leq l \leq m$, and for $n \geq 2$ two numbers are defined:

$$
\mu_{C_k,C_l} = 2\frac{S_{k,l}^{\mu}}{n(n-1)}, \quad \nu_{C_k,C_l} = 2\frac{S_{k,l}^{\nu}}{n(n-1)}.
$$

The pair constructed from these two numbers plays the role of the intuitionistic fuzzy evaluation of the relations that can be established between any two criteria C_k

and C_l. In this way the index matrix M that relates evaluated objects with evaluating criteria can be transformed to another index matrix M^* that gives the relations among the criteria:

$$M^* = \begin{array}{c|ccc} & C_1 & \cdots & C_m \\ \hline C_1 & \langle \mu_{C_1,C_1}, \nu_{C_1,C_1} \rangle & \cdots & \langle \mu_{C_1,C_m}, \nu_{C_1,C_m} \rangle \\ \vdots & \vdots & \ddots & \vdots \\ C_m & \langle \mu_{C_m,C_1}, \nu_{C_1,C_m} \rangle & \cdots & \langle \mu_{C_m,C_m}, \nu_{C_m,C_m} \rangle \end{array}$$

The final step of the algorithm is to determine the degrees of correlation between the criteria, depending on the user's choice of μ and ν. We call these correlations between the criteria: 'positive consonance,' 'negative consonance' or 'dissonance.'

Let α, β [0; 1] be given, so that $\alpha + \beta \leq 1$. We say that criteria C_k and C_l are in:

- (α, β)-positive consonance, if $\mu_{C_k,C_l} > \alpha$ and $\nu_{C_k,C_l} < \beta$;
- (α, β)-negative consonance, if $\mu_{C_k,C_l} < \beta$ and $\nu_{C_k,C_l} > \alpha$;
- (α, β)-dissonance, otherwise, [1].

Obviously, the larger α and/or the smaller β, the less number of criteria may be simultaneously connected with the relation of (α, β)-positive consonance. For practical purposes, it carries the most information when either the positive or the negative consonance is as large as possible, while the cases of dissonance are less to no informative.

3   Defining the ICA Thresholds: Overview

According to the authors, one of the most important aspects of the method is the way the thresholds are determined: as this is what gives the measure of precision of the decision made using ICA. Setting the thresholds can, obviously, be a completely expert's decision, but an algorithmic approach is considered to yield more precision, objectivity and sustainability. This is why a series of papers has explored this issue in a stepwise manner, based on both our deepening understanding of the ICA method, and observations made during the application of the method to various real-life datasets. It is noteworthy that from the very first contemplations of the method, we realized that although both the values of the intercriteria consonance, and respectively the thresholds against these being measured, are normed within the [0, 1]-interval, the thresholds are not universal, and different thresholds may be adequate in different problem areas of application. In other words, there is no "one size fits all" solution when defining the ICA thresholds and it is worth specifically exploring how ICA membership and non-membership thresholds α and β can be objectively defined.

Historically, "*the immediate first idea*" (per [26]) of this leg of research involved setting predefined values, i.e. numbers from the [0, 1]-interval, relatively high for

Table 1 Results from application of ICA over a dataset of EU28 member states' competitiveness in year 2013–2014, based on the data from [28]

μ	1	2	3	4	5	6	7	8	9	10	11	12
1	1.000	0.735	0.577	0.720	0.807	0.836	0.733	0.749	0.854	0.503	0.804	0.844
2	0.735	1.000	0.479	0.661	0.749	0.677	0.537	0.590	0.786	0.661	0.804	0.799
3	0.577	0.479	1.000	0.421	0.519	0.558	0.627	0.675	0.550	0.413	0.548	0.556
4	0.720	0.661	0.421	1.000	0.730	0.683	0.590	0.563	0.677	0.497	0.712	0.690
5	0.807	0.749	0.519	0.730	1.000	0.735	0.622	0.632	0.775	0.579	0.815	0.847
6	0.836	0.677	0.558	0.683	0.735	1.000	0.749	0.712	0.788	0.466	0.759	0.751
7	0.733	0.537	0.627	0.590	0.622	0.749	1.000	0.741	0.685	0.399	0.624	0.624
8	0.749	0.590	0.675	0.563	0.632	0.712	0.741	1.000	0.712	0.497	0.688	0.680
9	0.854	0.786	0.550	0.677	0.775	0.788	0.685	0.712	1.000	0.526	0.810	0.831
10	0.503	0.661	0.413	0.497	0.579	0.466	0.399	0.497	0.526	1.000	0.611	0.598
11	0.804	0.804	0.548	0.712	0.815	0.759	0.624	0.688	0.810	0.611	1.000	0.873
12	0.844	0.799	0.556	0.690	0.847	0.751	0.624	0.680	0.831	0.598	0.873	1.000

ν	1	2	3	4	5	6	7	8	9	10	11	12
1	0.000	0.220	0.386	0.188	0.132	0.077	0.185	0.172	0.090	0.452	0.138	0.111
2	0.220	0.000	0.466	0.228	0.172	0.228	0.362	0.317	0.146	0.286	0.135	0.138
3	0.386	0.466	0.000	0.476	0.405	0.344	0.286	0.251	0.394	0.537	0.394	0.389
4	0.188	0.228	0.476	0.000	0.143	0.169	0.283	0.307	0.201	0.397	0.175	0.198
5	0.132	0.172	0.405	0.143	0.000	0.153	0.272	0.259	0.135	0.341	0.098	0.079
6	0.077	0.228	0.344	0.169	0.153	0.000	0.135	0.169	0.101	0.439	0.143	0.159
7	0.185	0.362	0.286	0.283	0.272	0.135	0.000	0.146	0.209	0.505	0.267	0.275
8	0.172	0.317	0.251	0.307	0.259	0.169	0.146	0.000	0.206	0.415	0.217	0.233
9	0.090	0.146	0.394	0.201	0.135	0.101	0.209	0.206	0.000	0.405	0.119	0.101
10	0.452	0.286	0.537	0.397	0.341	0.439	0.505	0.415	0.405	0.000	0.328	0.344
11	0.138	0.135	0.394	0.175	0.098	0.143	0.267	0.217	0.119	0.328	0.000	0.071
12	0.111	0.138	0.389	0.198	0.079	0.159	0.275	0.233	0.101	0.344	0.071	0.000

(a) Membership parts of the IF pairs   (b) Non-membership parts of the IF pairs

the membership threshold, like numbers in the interval [0.75, 1], and relatively low for the non-membership threshold β like numbers in the interval [0, 0.25]. The research (published in paper [27]) started from the ideal membership threshold of 1 and non-membership threshold of 0, and involved gradual decrease of α and increase of, checking on each step what new intercriteria pairs "emerge." Checking the ICA pairs against both thresholds was consequent, with first checking the emergent ICA pairs when applying the membership threshold, and then the non-membership one.

In [27], we presented the essence of the idea, applying the method to data from the Global Competitiveness Report of the World Economic Forum [28], where we were interested to detect the eventual correlations between the 12 'pillars of country competitiveness,' in order to outline those fewer pillars on which policy makers should concentrate their efforts. Our motivation to conduct the analysis has been that it might be expected that improved country's performance against some pillars would positively affect the country's performance in the respective correlating ones. This is in line with WEF's address to state policy makers to "*identify and strengthen the transformative forces that will drive future economic growth*" of the countries, as formulated in the Preface of the GCR in the year 2013–2014 [28].

The results from application of the ICA method over the data from this dataset is given in the two tables below (Table 1a, b), where for each pair of criteria $C_i - C_j$ the degree of correlation (consonance) is given by the intuitionistic fuzzy (IF) pair $\langle \mu_{C_i,C_j}, \nu_{C_i,C_j} \rangle$. For every criterion, the degree of correlation $C_i - C_i$ is $\langle 1, 0 \rangle$, because every criterion perfectly correlates only with itself; also the tables are symmetrical along the main diagonal, because the degree of correlation of $C_i - C_j$ equals that of $C_j - C_i$.

In [27], we showed how, given a fixed year (i.e. dataset), we variate the thresholds α, β, which respectively changes the number of criteria that start correlating, hence, the positive consonances formed between the pairs of criteria. As an illustration, we

Table 2 List of positive intercriteria consonances per different pairs of thresholds

Values for $(\alpha; \beta)$	List of positive consonances (PC)	No. of PC	No. of criteria
(0.85; 0.15)	**1–9; 11–12**	2	4
	1–5; 1–6; 1–9; **1–11; 1–12; 2–11; 5–11;** **5–12; 9–11; 9–12;** 11–12	11	7
(0.75; 0.25)	1–5; 1–6; 1–9; 1–11; 1–12; **2–9;** 2–11; **2–12; 5–9;** 5–11; 5–12; **6–9; 6–11; 6–12;** 9–11; 9–12; 11–12	17	7
(0.70; 0.30)	**1–2; 1–4;** 1–5; 1–6; **1–7; 1–8;** 1–9; 1–11; 1–12; **2–5;** 2–9; 2–11; 2–12; **4–5; 4–11;** **5–6;** 5–9; 5–11; 5–12; **6–7; 6–8;** 6–9; 6–11; 6–12; **7–8; 8–9;** 9–11; 9–12; 11–12	29	10
(0.65; 0.35)	1–2; 1–4; 1–5; 1–6; 1–7; 1–8; 1–9; 1–11; 1–12; **2–4;** 2–5; **2–6;** 2–9; **2–10;** 2–11; 2–12; **3–8;** 4–5; **4–6; 4–9;** 4–11; **4–12;** 5–6; 5–9; 5–11; 5–12; 6–7; 6–8; 6–9; 6–11; 6–12; 7–8; **7–9;** 8–9; **8–11; 8–12;** 9–11; 9–12; 11–12	39	12

give here (Table 2) the results from the check of the dataset of EU28 countries from the GCR in the year 2013–2014 how stepwise changing the thresholds α, β with step of 0.05 leads to "emergence" of new intercriteria positive consonances and new correlating criteria (those highlighted with bold typeface).

In the paper [27], a finer step, with which the thresholds α, β are changed, was taken, namely, 0.025 (decrease for α, increase for β) starting with (0.85, 0.15) and ending with (0.7, 0.3). The observations of the results lead us to the conclusions that in the low end, under a certain value for threshold α (respectively, above a certain value for threshold β), it is natural that *all* criteria start exhibiting some (rather low) degree of consonance, which is ineffective for the analysis. In the high end, with too high threshold α, respectively too threshold β, so few, if any, criteria are in positive consonance that this is also not effective either. Thus, specifically for the case of the global competitiveness of the EU28 countries, we concluded to focus the analysis of intercriteria consonances in the thresholds range from (0.775; 0.225) to (0.825; 0.175), with some more specific discussions about the findings being made.

In [29], already it was noted that this approach may yield some rather different results for α and β. It was commented that paper [27] adopted the "*simplistic case*," where the values of the thresholds α and β in the different pairs of numbers were always summing up to 1, like (0.85; 0.15), (0.825; 0.175), (0.8; 0.2), etc. This however helped us could notice that with this setting applied over the data in the dataset, threshold β produces disproportionately many intercriteria pairs, far beyond the number of pairs produced by the respective threshold α. As we summarized in Table 4 in [30], when $\alpha = 0.85$, the outlined pairs in positive consonance are only 2, and when $\beta = 1 - \alpha = 0.15$, the outlined pairs in negative consonance are 19. When $(\alpha, \beta) = (0.80; 0.20)$, these numbers are respectively 11 and 29; and so forth, using

Table 3 Finding maximal correlations per criterion	Column (1)	Column (2)
	C_1	$\max\limits_{j, j \neq 1} \mu(C_1, C_j)$
	C_2	$\max\limits_{j, j \neq 2} \mu(C_2, C_j)$
	...	...
	C_n	$\max\limits_{j, j \neq n} \mu(C_n, C_j)$

a step of 0.05, until we reach the intuitionistic fuzzy pair of thresholds (0.65; 0.35), when these numbers are respectively 39 and 51.

This led to the idea of a finer approach in which one of the thresholds, interchangeably, was fixed, and the other one was to be accordingly determined, in order to yield commensurate (if not identical) results. This required redefinition of the problem of determining the thresholds: it was no more the question to determine the exact values of the thresholds α and β, and thus obtain which pairs of criteria exhibit high positive consonance, but it was the reverse problem: in order to obtain the values of both thresholds it was necessary to determine a number $0 < k < n$ and find which are those k (out of all n) criteria, which exhibit the highest positive intercriteria consonances.

For this purpose, in [26] a simple algorithm was developed, which for every separate criterion calculates its positive consonance with each of the rest $(n - 1)$ criteria, and takes the maximal value; and then sorts these n values in a descending way, selecting the top k of them. However, it was noted that following such an algorithm it is quite possible to "draw the line" at a wrong place: the differences between the α-s and β-s of the first $(k + 1)$ criteria and between the α-s and β-s of the first k criteria might be negligible, while the differences between the α-s and β-s of the first $(k - 1)$ criteria and between the α-s and β-s of the first k criteria might be quite well expressed. Later, this motivated the discussion in [31], aimed to prescind from a particular number k, but attempt to identify the (unknown in advance) "most strongly correlating criteria."

The algorithm, developed in [26] is the following.

(1) For each criterion C_i, $i = 1, ..., n$, we find $\max\limits_{j, j \neq i} \mu(C_i, C_j)$, i.e. the maximum of the discovered correlations of C_i with all the rest criteria $C_j, j = 1, ..., n, i \neq j$. Thus we obtain for each criterion, which is its top-correlating value.
(2) We create a table like the one shown on Table 3.
(3) We sort the whole Table 3 by Column (2) in descending order, thus ordering the top-correlating values for all individual criteria.
(4) We shortlist the first k criteria from Column (1) in the resultant sorted table.
(5) The sought value of the threshold α is then the respective value in Column (2) on k-th place top down, in the resultant sorted table.

The algorithm is shown with the following example, which also illustrates why "mechanically" determining the number of k criteria, we would like to work with, is not obligatory producing optimal results: the gap between the value of α, needed to shortlist the top 4 correlating criteria (0.854) and α, needed to shortlist the top 5 correlating criteria (0.847) is much smaller compared to the gap between the value α, needed to shortlist the top 5 and that, needed to shortlist the top 6 correlating criteria (0.804), or the gap between the value α, needed to shortlist the top 3 and that, needed to shortlist the top 4 correlating criteria (0.873) (Fig. 1).

After sorting the column by $\max_{j,j\neq i} \mu(C_i, C_j)$ in descending order, we obtain the table from Fig. 2.

This concluded the algorithm for determining the membership threshold α. In order to determine the respective value for the non-membership threshold β, we repeat the algorithm for β in a mirror-like way: we find $\min_{j,j\neq i} \nu(C_i, C_j)$, i.e. the minimums of the discovered correlations of C_i with all the rest criteria $C_j, j = 1, \ldots, n, i \neq j$, and then sort them in ascending order. Then we take the *minimal superset of criteria* ordered by negative consonance, which contains the set of criteria as defined by the number k in the algorithm for defining threshold α, as graphically shown on Fig. 3.

In our example, the set of criteria ordered by positive consonance with $k = 4$ was the set {11, 12, 1, 9} when $\alpha = 0.854$, hence the resultant minimal subset from the set of criteria ordered by positive consonance is {11, 12, 1, 6, 5, 9} when $\beta = 0.09$. We can note that in this way the sum of α and β is no more 1. In comparison with the former version of the method of determining the thresholds by taking predefined constants summing up to 1, this would mean that if β was set to $1 - 0.854 = 0.146$, this would mean effectively that from the non-membership side 10 (rather than 6) out

μ	1	2	3	4	5	6	7	8	9	10	11	12
1		0.735	0.577	0.720	0.807	0.836	0.733	0.749	0.854	0.503	0.804	0.844
2	0.735		0.479	0.661	0.749	0.677	0.537	0.590	0.786	0.661	0.804	0.799
3	0.577	0.479		0.421	0.519	0.558	0.627	0.675	0.550	0.413	0.548	0.556
4	0.720	0.661	0.421		0.730	0.683	0.590	0.563	0.677	0.497	0.712	0.690
5	0.807	0.749	0.519	0.730		0.735	0.622	0.632	0.775	0.579	0.815	0.847
6	0.836	0.677	0.558	0.683	0.735		0.749	0.712	0.788	0.466	0.759	0.751
7	0.733	0.537	0.627	0.590	0.622	0.749		0.741	0.685	0.399	0.624	0.624
8	0.749	0.590	0.675	0.563	0.632	0.712	0.741		0.712	0.497	0.688	0.680
9	0.854	0.786	0.550	0.677	0.775	0.788	0.685	0.712		0.526	0.810	0.831
10	0.503	0.661	0.413	0.497	0.579	0.466	0.399	0.497	0.526		0.611	0.598
11	0.804	0.804	0.548	0.712	0.815	0.759	0.624	0.688	0.810	0.611		0.873
12	0.844	0.799	0.556	0.690	0.847	0.751	0.624	0.680	0.831	0.598	0.873	

C_i	$\max_{j,j\neq i} \mu(C_i, C_j)$
1	0.854
2	0.804
3	0.675
4	0.730
5	0.847
6	0.836
7	0.749
8	0.749
9	0.854
10	0.661
11	0.873
12	0.873

Fig. 1 Finding maximal correlations per criterion (Step 2 of the algorithm)

C_i	$\max\limits_{j,j\neq i} \mu(C_i,C_j)$
11	0.873
12	0.873
1	0.854
9	0.854
5	0.847
6	0.836
2	0.804
7	0.749
8	0.749
4	0.73
3	0.675
10	0.661

$k = 4$

Number of correlating criteria	Number of pairs of correlating criteria	Criteria ordered by positive consonance	True when $\alpha \geq$
2	1	11	0.873
		12	
4	2	1	0.854
		9	
5	3	5	0.847
6	11	2	0.804
7	13	6	0.788
9	20	7	0.749
		8	
10	25	4	0.730
11	37	3	0.675
12	39	10	0.661

Fig. 2 Shortlist the first k criteria (Step 4 of the algorithm)

$k = 4$

Number of correlating criteria	Number of pairs of correlating criteria	Criteria ordered by positive consonance	True when $\alpha \geq$
2	1	11	0.873
		12	
4	2	1	0.854
		9	
5	3	5	0.847
6	11	2	0.804
7	13	6	0.788
9	20	7	0.749
		8	
10	25	4	0.730
11	37	3	0.675
12	39	10	0.661

Number of correlating criteria	Number of pairs of correlating criteria	Criteria ordered by negative consonance	True when $\beta \leq$
2	1	11	0.071
		12	
4	2	1	0.077
		6	
5	3	5	0.079
6	4	9	0.09
8	13	2	0.135
		7	
9	17	4	0.143
10	19	8	0.146
11	38	3	0.251
12	45	10	0.286

Fig. 3 Determining the appropriate threshold β with respect to threshold α

of the 12 criteria would be interacting within this threshold, which is much bigger "noise."

This "noise" resulting from the consecutive processing of the membership and non-membership thresholds was observed and solved in the next publication [32] with a solution inspired by the triangular geometric interpretation of intuitionistic fuzzy sets. The intuitionistic fuzzy triangle (proposed for the first time in [33], also see [11]) is a graphic interpretation of intuitionistic fuzzy sets that has no analogue in ordinary fuzzy sets. Its vertices (1, 0), (0, 1) and (0, 0) represent respectively the absolute Truth, the absolute Falsity and the absolute Uncertainty, while the hypotenuse is a projection of the intuitionistic fuzzy onto the fuzzy.

Fig. 4 Close-up of
shortlisted intercriteria pairs
in the IF triangle cut-out

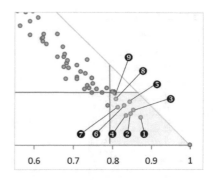

Plotting the intercriteria pairs as points onto the intuitionistic fuzzy interpretational triangle helped for the first time to develop a procedure for short-listing the set of top consonance pairs of criteria according to both thresholds α and β simultaneously. For this purpose, for each point its distance from the $(1, 0)$ point is calculated, where $(1, 0)$ in the context of ICA represents the case of perfect positive consonance between two criteria (including the perfect positive consonance of any criterion with itself).

The formula for the distance d_{C_i, C_j} of the intercriteria pair (C_i, C_j) to the $(1; 0)$ point is

$$d_{C_i, C_j} = \sqrt{\left(1 - \mu_{C_i, C_j}\right)^2 + v_{C_i, C_j}^2}$$

and the pairs are ordered according to their d_{C_i, C_j} sorted in ascending way. Plotting the intercriteria correlations as points in the intuitionistic fuzzy triangle gives us the possibility to rank and work with the strongest pairs of criteria, simultaneously handling membership and non-membership at a time, while in the hitherto steps of our research we have ranked and worked with the individual criteria, ordered according to one of the components in the pair, most often the membership component. In addition, this innovative approach to the theory of ICA led to a new feature in the software for calculation of intercriteria correlations, enabling their embedded graphical visualization [8].

To demonstrate the advantage of this approach, we gave in [26] the same data for the competitiveness of EU28 economies from the World Economic Forum's Global Competitiveness Report for the year 2013–2014. For convenience, the ICA software arranges the output of intercriteria consonances in two tables, one giving the membership parts (Table 4) and the other giving the non-membership parts (Table 5) of the intuitionistic fuzzy pairs for each pair of criteria. Then we give the graphic interpretation (Fig. 4) of the so-produced intuitionistic fuzzy set of points (standing for the intercriteria pairs, 66 in number for the case of 12 criteria). We provide also (the beginning of) the table (Table 6) with ordered correlating pairs, with respect to the distance from the $(1, 0)$ point.

To define the top correlating criteria, as explained in [26], two randomly taken threshold values (namely, $\alpha = 0.796$ and $\beta = 0.134$) are taken, which yield 9 points

Table 4 Discovered membership values for the year 2013–2014

M^μ	1	2	3	4	5	6	7	8	9	10	11	12
1		0.74	0.58	0.72	0.81	0.84	0.73	0.75	0.85	0.5	0.8	0.84
2	0.74		0.48	0.66	0.75	0.68	0.54	0.59	0.79	0.66	0.8	0.8
3	0.58	0.48		0.42	0.52	0.56	0.63	0.67	0.55	0.41	0.55	0.56
4	0.72	0.66	0.42		0.73	0.68	0.59	0.56	0.68	0.5	0.71	0.69
5	0.81	0.75	0.52	0.73		0.74	0.62	0.63	0.78	0.58	0.81	0.85
6	0.84	0.68	0.56	0.68	0.74		0.75	0.71	0.79	0.47	0.76	0.75
7	0.73	0.54	0.63	0.59	0.62	0.75		0.74	0.69	0.4	0.62	0.62
8	0.75	0.59	0.67	0.56	0.63	0.71	0.74		0.71	0.5	0.69	0.68
9	0.85	0.79	0.55	0.68	0.78	0.79	0.69	0.71		0.53	0.81	0.83
10	0.5	0.66	0.41	0.5	0.58	0.47	0.4	0.5	0.53		0.61	0.6
11	0.8	0.8	0.55	0.71	0.81	0.76	0.62	0.69	0.81	0.61		0.87
12	0.84	0.8	0.56	0.69	0.85	0.75	0.62	0.68	0.83	0.6	0.87	

Table 5 Discovered non-membership values for the year 2013–2014

M^ν	1	2	3	4	5	6	7	8	9	10	11	12
1		0.22	0.39	0.19	0.13	0.08	0.19	0.17	0.09	0.45	0.14	0.11
2	0.22		0.47	0.23	0.17	0.23	0.36	0.32	0.15	0.29	0.13	0.14
3	0.39	0.47		0.48	0.4	0.34	0.29	0.25	0.39	0.54	0.39	0.39
4	0.19	0.23	0.48		0.14	0.17	0.28	0.31	0.2	0.4	0.17	0.2
5	0.13	0.17	0.4	0.14		0.15	0.27	0.26	0.13	0.34	0.10	0.08
6	0.08	0.23	0.34	0.17	0.15		0.13	0.17	0.1	0.44	0.14	0.16
7	0.19	0.36	0.29	0.28	0.27	0.13		0.15	0.21	0.51	0.27	0.28
8	0.17	0.32	0.25	0.31	0.26	0.17	0.15		0.21	0.42	0.22	0.23
9	0.09	0.15	0.39	0.2	0.13	0.1	0.21	0.21		0.4	0.12	0.10
10	0.45	0.29	0.54	0.4	0.34	0.44	0.51	0.42	0.4		0.33	0.34
11	0.14	0.13	0.39	0.17	0.10	0.14	0.27	0.22	0.12	0.33		0.07
12	0.11	0.14	0.39	0.2	0.08	0.16	0.28	0.23	0.10	0.34	0.07	

in the cut-out, i.e. 9 intercriteria positive consonance pairs, formed among a set of 6 individual criteria. As we will see, there are also other alternative way of forming the subset of top correlating intercriteria pairs.

In a consequent step of this leg of research, we explored the question of traversing and ranking elements of an intuitionistic fuzzy set in the intuitionistic fuzzy interpretation triangle per their proximity to the point (1, 0). The procedure proposed in [34] required that these threshold values α, β are known in advance and predefined, which for various reasons may not always be the case. For this purpose, the procedure requires application of the topological operators *Closure* and *Interior*, which are defined using the following formulas (see [1, 22, 35]) and illustrated on Fig. 5:

Table 6 Ordering of the correlating pairs, with respect to the distance from (1, 0) of the points that represent them in the IF interpretational triangle

No.	d_{C_i,C_j}	Criteria in (α, β)-positive consonance	μ_{C_i,C_j}	ν_{C_i,C_j}
❶	0.148	**11–12** Business sophistication—Innovation	0.87	0.07
❷	0.170	**5–12** Higher education and training—Innovation	0.85	0.08
❸	0.175	**1–9** Institutions—Technological readiness	0.85	0.09
❹	0.179	**1–6** Institutions—Goods market efficiency	0.84	0.08
❺	0.194	**1–12** Institutions—Innovation	0.84	0.11
❻	0.197	**9–12** Technological readiness—Innovation	0.83	0.10
❼	0.206	**5–11** Higher education and training—Business sophistication	0.82	0.10
❽	0.225	**9–11** Technological readiness—Business sophistication	0.81	0.12
❾	0.230	**1–5** Institutions—Higher education and training	0.81	0.13

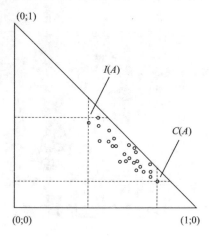

Fig. 5 An IFS, plotted onto the intuitionistic fuzzy triangle, with the indicated places of the topological operators Closure and Interior

$$C(A) = \left\{ \left\langle x, \sup_{y \in E} \mu_A(y), \inf_{y \in E} \nu_A(y) \right\rangle | x \in E \right\},$$

$$I(A) = \left\{ \left\langle x, \inf_{y \in E} \mu_A(y), \sup_{y \in E} \nu_A(y) \right\rangle | x \in E \right\}.$$

We will note that since, in the context of ICA, we are only working with finite sets of m objects, of n criteria, and therefore with a resultant finite set of $n(n-1)/2$

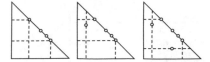

Fig. 6 Triangle, trapezium or pentagon are the possible shapes of the zone, enclosed by the topological operators Closure and Interior, and the hypotenuse of the intuitionistic fuzzy triangle

Fig. 7 The segment from the intuitionistic fuzzy triangle, containing the pentagonal zone, enclosed by the topological operators Closure and Interior (respectively, points R and P) and the hypotenuse, gridded with unit rectangles

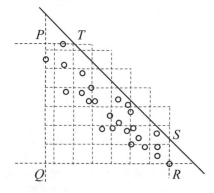

intercriteria pairs, we can safely replace the functions 'supremum' and 'infimum,' respectively by the functions 'maximum' and 'minimum.'

We will also note that depending on the particular set, it may have the form of a triangle in the case of a fuzzy set, all of which elements are plotted onto the hypotenuse (the least challenging case), or a trapezium, or a pentagon (the most challenging case), see Fig. 6. In our research in [34] and here we discuss the most general case of a pentagon (Fig. 7).

The procedure for ranking the intercriteria pairs, hence, comprises two phases: (1) To define the unit lengths of the rectangular grid that will divide the pentagon; (2) To define the consequence of traversing the so-defined subrectangles of the grid. For completeness of the present discussion, we will note that the starting point of the so-constructed grid depends on the specific problem formulation: if the problem requires us to seek highest possible consonances among the intercriteria pairs, i.e. those closest to the Truth, represented by point $(1, 0)$, the so-constructed grid is to start from the point (mapping) of the operator Closure (which for the needs of the example on Fig. 7 happens to be point R).

A possible way to define the unit lengths a, b of the rectangular grid is given with the following two formulas:

$$a = \frac{\max\limits_{y \in E} \mu_A(y) - \min\limits_{y \in E} \mu_A(y)}{\frac{n(n-1)}{2}}, \quad b = \frac{\max\limits_{y \in E} \nu_A(y) - \min\limits_{y \in E} \nu_A(y)}{\frac{n(n-1)}{2}}$$

The lengths PQ and QR are divided by the total number of points in the plotted set, and this is the finest possible division for the grid.

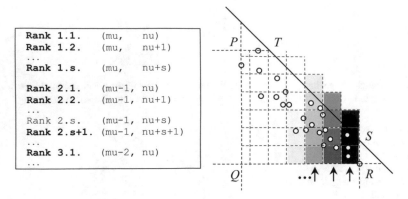

Rank 1.1.	(mu,	nu)
Rank 1.2.	(mu,	nu+1)
...		
Rank 1.s.	(mu,	nu+s)
Rank 2.1.	(mu-1,	nu)
Rank 2.2.	(mu-1,	nu+1)
...		
Rank 2.s.	(mu-1,	nu+s)
Rank 2.s+1.	(mu-1,	nu+s+1)
...		
Rank 3.1.	(mu-2,	nu)
...		

Fig. 8 Traversing the grid by the strategy "max μ first"

Another approach is to assign to a and b, respectively, the smallest possible positive, non-null difference in the first coordinates of any two points of the set, and the smallest possible positive, non-null difference in the second coordinates of any two points in the set, by the formulas:

$$a = \min_{i,j \in A}\left(|\mu_i - \mu_j|\right), \ b = \min_{k,l \in A}\left(|\nu_k - \nu_l|\right).$$

For the sake of completeness, we can also note the most obvious way of defining the unit lengths of the rectangular grid by dividing PQ and QR into predefined number(s), not necessarily the same number of sections per side. Then, for two predefined numbers u, w, the formulas will have the following forms:

$$a = \frac{\max_{y \in E} \mu_A(y) - \min_{y \in E} \mu_A(y)}{u}, \ b = \frac{\max_{y \in E} \nu_A(y) - \min_{y \in E} \nu_A(y)}{w}.$$

Here we make the assumption that all different intercriteria pairs whose mappings (points) that belong to a single subrectangle of the grid (if more than one) will be treated equally. This transforms the question to comparing and ranking the subrectangles of the grid, i.e., how the grid is being traversed, which essentially reduces to the question how we treat—and prioritize between—the three intuitionistic fuzzy components: membership, non-membership and uncertainty. In this relation, in [34] we proposed three different strategies, or scenarios.

(1) *Strategy "max μ first."* In response to this strategy, we start with the subrectangle with the maximal membership and minimal non-membership, i.e. the one containing the set's closure, and traverse through the grid in vertical direction (bottom-to-top), in a way that preserves the membership part as high as possible, while running through the gradually increasing non-membership parts, as illustrated in Fig. 8. The following pseudocode gives it in a more formal way:

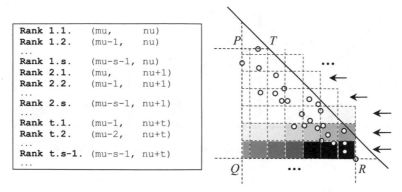

Rank 1.1.	(mu,	nu)
Rank 1.2.	(mu-1,	nu)
...		
Rank 1.s.	(mu-s-1,	nu)
Rank 2.1.	(mu,	nu+1)
Rank 2.2.	(mu-1,	nu+1)
...		
Rank 2.s.	(mu-s-1,	nu+1)
...		
Rank t.1.	(mu-1,	nu+t)
Rank t.2.	(mu-2,	nu+t)
...		
Rank t.s-1.	(mu-s-1,	nu+t)
...		

Fig. 9 Traversing the grid by the strategy "min ν first"

Rank 1.	(mu, nu)
Rank 2.	(mu-1, nu),
	(mu+1, nu)
Rank 3.	(mu-2, nu),
	(mu-1, nu+1),
	(mu, nu+2)
...	
Rank r+1.	(mu-r, nu),
	(mu-r+1, nu+1),
	(mu-r+2, nu+2),
	...,
	(mu-2, nu+r-2),
	(mu-1, nu+r-1),
	(mu, nu+r)
...	

Fig. 10 Traversing the grid by the diagonal strategy

(2) *Strategy "min ν first".* In response to this strategy, we again start with the subrectangle with the maximal membership and minimal non-membership, but this time we traverse through the grid in horizontal direction (right-to-left), in a way that preserves the non-membership part as low as possible, while running through the gradually decreasing membership parts. The illustration of this strategy follows by analogy, Fig. 9. The following pseudocode gives it in a more formal way:

(3) *Diagonal strategy.* Starting with the subrectangle with the maximal membership and minimal non-membership, we then take the simultaneously the union of the subrectangles that are located one up and one left of the previous, and so forth, as illustrated in Fig. 4 and by the following pseudocode (Fig. 10).

Any of the strategies for defining the rectangular grid can be combined with any strategy for traversing the grid; other alternatives are also possible.

In the latest development of the study of ICA thresholds determination, other alternatives were proposed about how to form the subset of top correlating intercriteria pairs. In [36], three alternatives were listed as ways to construct that subset,

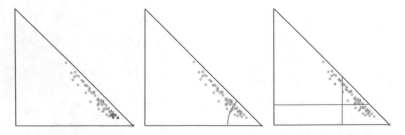

Fig. 11 Illustrating the three alternatives for constructing the subset of ICA pairs

depending on user's preference or external requirement, and a speculation was made that others are also possible. There three proposed alternatives were graphically represented as in Fig. 11, and defined as follows:

(1) Select top p or top $q\%$ of the $n(n-1)/2$ ICA pairs;
(2) Select all ICA pairs whose corresponding points are within a given radius r from the $(1; 0)$ point;
(3) Select all ICA pairs whose corresponding points fall within the trapezoid formed between the abscissa, the hypotenuse and the two lines corresponding to $y = \alpha$, $x = \beta$ for two predefined numbers $\alpha, \in [0; 1]$ (i.e., the approach adopted earlier in [26]).

Most notably, this leg of the research in [36, 37] was dedicated to idea of having *triples* of criteria in positive consonance, upgrading the original concept for discovery of consonances between intercriteria pairs. The formulated work hypothesis was that, given a record of intercriteria pairs that have exhibited positive consonance over a longer period of time, triples and n-tuples of more criteria could be detected among them featuring high enough pairwise consonance, and an algorithm was proposed for identification and ranking of the intercriteria triples. The particular interpretation of such triple of intercriteria consonances, as well as the practical usefulness of the new concept, was discussed to be a matter of further investigation by problem-specific experts.

4   Conclusion

In the present paper, we aimed to make a detailed and justified overview of the stages of development of the research of defining the thresholds in intercriteria analysis. Starting with the definition and the first intuitive steps related to determining the thresholds, we trace its progress over time based on our deepening understanding of the ICA method and following the practical applications of ICA over diverse problems and datasets. We consider this leg of the ICA research to be one of the most important theoretical aspects, since it is the concluding step of ICA, determining the measure of precision which ICA-based decisions can be made with.

Nevertheless, we consider it appropriate other case studies and problems, approached in future with the ICA approach, all proposed algorithms—from the present and previous researches—are worth approbating, in order to compare the results for various fields of application, consult them with experts in the respective areas, and make a better justification of our choice of method of selecting threshold values and selecting the top-correlating criteria.

Acknowledgements The authors are grateful for the support under the Grant Ref. No. DFNI-I-02-5 "Intercriteria Analysis: A Novel Method for Decision Making" funded by the Bulgarian National Science Fund.

References

1. K. Atanassov, D. Mavrov, V. Atanassova. Intercriteria Decision Making: A New Approach for Multicriteria Decision Making, Based on Index Matrices and Intuitionistic Fuzzy Sets. Issues in Intuitionistic Fuzzy Sets and Generalized Nets, Vol. 11, 2014, 1–8.
2. K. Atanassov, V. Atanassova, G. Gluhchev, InterCriteria Analysis: Ideas and problems, Notes on Intuitionistic Fuzzy Sets, Vol. 21, 2015, No. 1, 81–88.
3. L. Todorova, P. Vassilev, J. Surchev. Using Phi Coefficient to Interpret Results Obtained by InterCriteria Analysis, In: Novel Developments in Uncertainty Representation and Processing, Vol. 401, Advances in Intelligent Systems and Computing, Springer, 2016, 231–239.
4. N. Angelova, K. Atanassov, B. Riecan, Intercriteria analysis of the intuitionistic fuzzy implication properties, Notes on Intuitionistic Fuzzy Sets, Vol. 21(5), 2015, 20–23.
5. P. Vassilev, L. Todorova, V. Andonov. An auxiliary technique for InterCriteria Analysis via a three dimensional index matrix. Notes on Intuitionistic Fuzzy Sets, 21(2), 2015, 71–76.
6. V. Traneva, V. Internal operations over 3-dimensional extended index matrices, Proceedings of the Jangjeon Mathematical Society, Vol. 18(4), 2015, 547–569.
7. D. Mavrov. Software for InterCriteria Analysis: Implementation of the main algorithm, Notes on Intuitionistic Fuzzy Sets, Vol. 21(2), 2015, 77–86.
8. D. Mavrov, I. Radeva, K. Atanassov, L. Doukovska, I. Kalaykov, InterCriteria Software Design: Graphic Interpretation within the Intuitionistic Fuzzy Triangle, Proceedings of the Fifth International Symposium on Business Modeling and Software Design, 2015, 279–283.
9. D. Stratiev, I. K Shishkova, A. Nedelchev, K. E. Kirilov, E. Nikolaychuk, A. S. Ivanov, I. Sharafutdinov, A. Veli, M. Mitkova, T. Tsaneva, N. Petkova, R. Sharpe, D. Yordanov, Z. Belchev, S. Nenov, N. Rudnev, V. Atanassova, E. Sotirova, S. Sotirov, K. Atanassov, Investigation of relationships between petroleum properties and their impact on crude oil compatibility, Energy & Fuels, American Chemical Society, 2015.
10. D. Stratiev, S. Sotirov, I. Shishkova, A. Nedelchev, I. Sharafutdinov, A. Vely, M. Mitkova, D. Yordanov, E. Sotirova, V. Atanassova, K. Atanassov, D. D. Stratiev, N. Rudnev, S. Ribagin Investigation of relationships between bulk properties and fraction properties of crude oils by application of the intercriteria analysis, Petroleum Science and Technology, Vol. 34, Issue 13, 2016, 1113–1120.
11. K. Atanassov. Intuitionistic Fuzzy Sets: Theory and Applications. Springer Physica-Verlag, Heidelberg, 1999.
12. S. Sotirov, V. Atanassova, E. Sotirova, V. Bureva, D. Mavrov, Application of the Intuitionistic Fuzzy InterCriteria Analysis Method to a Neural Network Preprocessing Procedure, 9th Conference of the European Society for Fuzzy Logic and Technology (EUSFLAT), 30.06-03.07.2015, Gijon, Spain, 2015, 1559–1564.
13. T. Ilkova, M. Petrov, Application of InterCriteria Analysis to the Mesta River Pollution Modelling, Notes on Intuitionistic Fuzzy Sets, Vol. 21(2), 2015, 118–125.

14. T. Ilkova, O. Roeva, P. Vassilev, M. Petrov, InterCriteria Analysis in Structural and Parameter Identification of L-lysine Production Model, Issues in Intuitionistic Fuzzy Sets and Generalized Nets, Vol. 12, 2015, 39–52.
15. M. Angelova, O. Roeva, T. Pencheva, InterCriteria Analysis of Crossover and Mutation Rates Relations in Simple Genetic Algorithm, Annals of Computer Science and Information Systems, Vol. 5, 2015, 419–424.
16. O. Roeva, O., P. Vassilev, InterCriteria Analysis of Generation Gap Influence on Genetic Algorithms Performance, In: Novel Developments in Uncertainty Representation and Processing, Part V, (K. T. Atanassov, O. Castillo, J. Kacprzyk, M. Krawczak, P. Melin, S. Sotirov, E. Sotirova, E. Szmidt, G. De Tré, S. Zadrożny, Eds.), Vol. 401, Advances in Intelligent Systems and Computing, 2016, 301–313.
17. O. Roeva, S. Fidanova, M. Paprzycki, InterCriteria Analysis of ACO and GA Hybrid Algorithms, Studies in Computational Intelligence, Springer, Vol. 610, 2016, 107–126.
18. S. Fidanova, O. Roeva, M. Paprzycki, InterCriteria Analysis of Ant Colony Optimization Application to GPS Surveying Problems, Issues in Intuitionistic Fuzzy Sets and Generalized Nets, Vol. 12, 2015, 20–38.
19. T. Pencheva, M. Angelova, P. Vassilev, O. Roeva, InterCriteria Analysis Approach to Parameter Identification of a Fermentation Process Model, In: Novel Developments in Uncertainty Representation and Processing, Part V, (K. T. Atanassov, O. Castillo, J. Kacprzyk, M. Krawczak, P. Melin, S. Sotirov, E. Sotirova, E. Szmidt, G. De Tré, S. Zadrożny, Eds.), Vol. 401, Advances in Intelligent Systems and Computing, 2016, 385–397.
20. K. Atanassov. Intuitionistic fuzzy sets, VII ITKR's Session, Sofia, June 1983 (in Bulgarian).
21. K. Atanassov. Intuitionistic fuzzy sets. Fuzzy Sets and Systems. Vol. 20 (1), 1986, 87–96.
22. K. Atanassov. On Intuitionistic Fuzzy Sets Theory. Springer, Berlin, 2012.
23. L. A. Zadeh. Fuzzy Sets. Information and Control Vol. 8, 1965, 333–353.
24. K. Atanassov. Generalized Nets. World Scientific, Singapore, 1991.
25. K. Atanassov. Index Matrices: Towards an Augmented Matrix Calculus. Springer, Cham, 2014.
26. V. Atanassova, I. Vardeva. Sum- and average-based approach to criteria shortlisting in the InterCriteria Analysis. Int. J. Notes on Intuitionistic Fuzzy Sets, Volume 20(4), 2014, 41–46.
27. V. Atanassova, L. Doukovska, D. Karastoyanov, F. Capkovic. InterCriteria Decision Making Approach to EU Member States Competitiveness Analysis: Trend Analysis. P. Angelov et al. (eds.), Intelligent Systems'2014, Advances in Intelligent Systems and Computing 322, 107–115.
28. World Economic Forum. The Global Competitiveness Reports, 2008–2017. http://www.weforum.org/issues/global-competitiveness.
29. V. Atanassova, D. Mavrov, L. Doukovska, K. Atanassov. Discussion on the threshold values in the InterCriteria Decision Making approach. International Journal Notes on Intuitionistic Fuzzy Sets, Volume 20(2), 2014, 94–99.
30. V. Atanassova, L. Doukovska, K. Atanassov, D. Mavrov – InterCriteria Decision Making Approach to EU Member States Competitiveness Analysis, Proc. of the International Symposium on Business Modeling and Software Design – BMSD'14, 24–26 June 2014, Lux-embourg, Grand Duchy of Luxembourg, 2014, 289–294.
31. V. Atanassova, L. Doukovska, D. Mavrov, K. Atanassov. InterCriteria Decision Making Approach to EU Member States Competitiveness Analysis: Temporal and Threshold Analysis. P. Angelov et al. (eds.), Intelligent Systems'2014, Advances in Intelligent Systems and Computing 322, 95–106.
32. V. Atanassova. Interpretation in the Intuitionistic Fuzzy Triangle of the Results, Obtained by the InterCriteria Analysis, 16th World Congress of the International Fuzzy Systems Association (IFSA), 9th Conference of the European Society for Fuzzy Logic and Technology (EUSFLAT), 30. 06-03. 07. 2015, Gijon, Spain, 2015, 1369–1374.
33. K. Atanassov. Geometrical Interpretation of the Elements of the Intuitionistic Fuzzy Objects, Mathematical Foundations of Artificial Intelligence Seminar, Sofia, 1989, Preprint IM-MFAIS-1-89. Reprinted: Int. J. Bioautomation, 2015, 19(4), Suppl. 2, S156–S171.

34. V. Atanassova, I. Vardeva, E. Sotirova, L. Doukovska. Traversing and Ranking of Elements of an Intuitionistic Fuzzy Set in the Intuitionistic Fuzzy Interpretation Triangle. In: Atanassov K. et al. (eds) Novel Developments in Uncertainty Representation and Processing. Advances in Intelligent Systems and Computing, Vol. 401. Springer, Cham, 2016.

35. K. Atanassov. Modal and topological operators, defined over intuitionistic fuzzy sets. In: Youth Scientific Contributions, Academic Publ. House, Sofia, 1, 1985, 18–21.

36. V. Atanassova, L. Doukovska, A. Michalikova, I. Radeva. Intercriteria analysis: From pairs to triples. Notes on Intuitionistic Fuzzy Sets, Vol. 22(5), 2016, 17–24.

37. L. Doukovska, V. Atanassova, G. Shahpazov, F. Capkovic, InterCriteria Analysis Applied to EU Micro, Small, Medium and Large Enterprises. In Proceedings of the Fifth International Symposium on Business Modeling and Software Design, Milan, Italy, 2015, 284–291.

Design and Comparison of ECG Arrhythmias Classifiers Using Discrete Wavelet Transform, Neural Network and Principal Component Analysis

Seyed Saleh Mohseni and Vahid Khorsand

Abstract Automatic classification of heartbeat is getting a significant value in today's medical systems. By implementation of these methods in portable diagnosis devices, many mortal diseases can be realized and cured in primary steps. In this paper two separate classifiers are designed and compared for heartbeat classification. The first strategy profits principal component analysis for feature extraction and neural network for classification whereas the second strategy utilizes discrete wavelet transform (DWT) for feature extraction and neural network (NN) as classifier. The arrhythmias which are investigated here include: normal beats (N), right bundle branch block (RBBB), left bundle branch block (LBBB), ventricular premature contraction (VPC) and paced beat (P). In addition, an output for unspecified signals is considered which devotes to anonymous signals which are not in the above list. The results show that both methods could achieve above 98% accuracy on MIT-BIH database.

1 Introduction

Every year many people die due to heart attacks. Early diagnosis of heart malfunction can prevent a lot of risky situations. For this reason, many portable devices such as holter-ECG are used to monitor the behavior of the cardiovascular system. As a result, an automatic system which can detect and alarm such a malfunction is of priority.

In recent years, many researches have been conducted in automatic classification of ECG signals. They are mainly different in the method of QRS complex detection, the type of utilized features, and the type of classifiers. Benali et al. [2] used a wavelet neural network for classification of five types of heartbeats and achieved more than 98% of accuracy for each category. They also used five features including three R–R features as well as QRS duration and R-amplitude for training the wavelet neural

S. S. Mohseni (✉) · V. Khorsand
Department of Electrical Engineering, Islamic Azad University, Nour Branch, Nour, Iran
e-mail: s_mohseni@iaunour.ac.ir; s_saleh_mohseni@yahoo.com

© Springer International Publishing AG, part of Springer Nature 2019
M Hadjiski and K T Atanassov (eds.), *Intuitionistic Fuzziness and Other Intelligent Theories and Their Applications*, Studies in Computational Intelligence 757,
https://doi.org/10.1007/978-3-319-78931-6_12

networks. In addition, different wavelets were evaluated as activation function of the neural network. Martis et al. [7] applied principal component analysis (PCA) on three different types of data i.e. the ECG signal, the prediction of ECG signal and the discrete wavelet transform (DWT) of ECG signal. Furthermore, they categorized the extracted features with different types of classifiers such as neural network and support vector machine (SVM). They achieved more than 98% accuracy in classifying five types of heartbeats. Martis et al. [6] used three dimensionality reduction techniques i.e. PCA, LDA and ICA on DWT sub-bands to extract features from five types of ECG signals. Moreover, they compared SVM, NN and probabilistic neural network (PNN) as classifiers on extracted features. Alickovi and Subasi [1] used multi-scale PCA for reducing the noise of the raw signal. Then they used DWT on ECG signal and extracted some statistical features from it. They also compared random forest, C4.5 and CART as classifiers and achieved their best result with random forest (more than 99% accuracy), in classification of five types of heartbeats. Thomas et al. [14] applied DWT and dual tree complex wavelet transform (DTCWT) on ECG signals and used the coefficients of wavelet transformations as well as some statistical information of ECG signal as features. Afterwards, they used multi-layer perceptron for classification. They achieved 91.23 and 94.64% sensitivity in test of five types of ECG signals. Sumathi et al. [13] used symlet wavelet transform on five types of ECG signals to extract the features. Then they used adaptive neuro-fuzzy inference system (ANFIS) as the classifier and achieved above 98% accuracy in distinguishing different classes. Jovic and Bogunovic [4] utilized some linear and nonlinear features of heart rate variation (HRV) as features and compared seven bunches of classifiers i.e. K-means, expectation maximization (EM), C4.5 decision tree, neural network, Bayesian network, SVM and random forest (RF). The classification was performed in two forms: four-class categorization and two-class (normal vs. abnormal) categorization. Ozbay [10] utilized complex discrete wavelet transform for feature extraction of eleven kinds of heartbeats and applied a complex valued artificial neural network as classifier. Mohseni and Mohammadyari [9] utilized a combination of PCA and HRV statistical features for classification of five types of heartbeats and the neural network as the classifier. Yu and Chen [16] utilized DWT with some statistical as well as frequency features plus a probabilistic neural network as classifier. Many other researches in the field of heartbeat classification can be found in Karimi Moridani et al. [5], Poorahangaryan et al. [12], Yeh et al. [15].

In this paper two different approaches for feature extraction from ECG signal are compared together. Then a neural network is applied on each group of feature to classify five types of heartbeats. The first method comprises PCA for dimensionality reduction of ECG data and neural network as classifier whereas the second method applies DWT as feature extractor and calculates the statistical features of DWT, then a neural network is utilized as classifier on extracted features. Five types of heartbeats which are classified here include: normal beats (N), right bundle branch block (RBBB), left bundle branch block (LBBB), ventricular premature contraction (VPC) and paced beat (P). All data are gathered from MIT-BIH Arrhythmia database [8]. Despite mentioned heartbeats, there are some other kinds of heartbeats which are rare but exist in each dataset. It includes paced beat and fusion of paced and normal beat. In this paper we categorized the other signals out of above five classes in a new class called anonymous class which comprises rare classes that exist in the dataset, i.e. atrial escape beat (A) and fusion of paced and normal beat (F). This new class was created to enable the machine to categorize undefined beats out of five mentioned classes. We omitted to classify the latter two mentioned classes in separate categories due to few numbers of existing samples in the database.

The rest of the paper is organized as follows. In Sect. 2, data extraction from MIT-BIH database, data gathering and preprocessing of data are discussed. Afterwards, in Sect. 3, two methods of feature extraction are surveyed. For this, Sect. 3.1 is devoted to PCA and Sect. 3.2 reviews the DWT. The type of classifier is also described in Sect. 3.3. Section 4 devotes to the simulation and results discussion. Finally Sect. 5 presents the conclusion.

2 Data Preparation

The data which is used in this paper is gathered from MIT-BIH Arrhythmia database. This database contains 48 records of ECG signal. The duration of each record is about half an hour. The sample time of digital records is 360 Hz and data are passed through a band-pass filter with frequency range of 0.1–100 Hz. Among records, 23 of them are chosen so that they show usual laboratory events and 25 records are chosen to contain some rare beats. In this website, ECG files are accompanied with an annotation. The annotation includes a label for each heartbeat. In other words, every record of ECG has a label on each beat. This label shows the type of heartbeat. Every record may consist of many types of beats. In this paper the records are categorized according to the annotations after some preprocessing.

The preprocessing stage utilizes Pan-Tompkins algorithm. Pan-Tompkins algorithm was firstly introduced in Pan and Tompkins [11] for detecting QRS complex. This algorithm was frequently used in many researches due to its high accuracy and robustness to noises. In Pan-Tompkins algorithm, firstly a band-pass filter which comprises a low-pass filter and a high-pass filter, is applied on the data to reduce the noises. Afterwards, the ECG signal is differentiated to highlight the high slope areas of the signal i.e. the QRS complex. Then, the result is squared which makes positive all the points and nonlinearly magnifies the higher frequencies of the differentiated signal. Next, a moving window integrates the signal inside the window. The goal of this stage is extracting the morphological information of the signal besides the rate of wave R. In the last step, a threshold is adjusted to detect the R waves and a decision is made on the detected peaks. Figure 1 shows the steps of Pan-Tompkins algorithm.

In MIT-BIH database, the sampling time of records is 360 Hz. So, a moving window of length 30 samples is chosen for the integration.

After detecting the QRS complexes, the ECG signals are segmented and categorized according to the annotation. The proper segmentation of data has an important role in the final results and may seriously decrease the quality of learning. For this, in this paper firstly the middle of two consecutive RRs is calculated then all data

Fig. 1 Pan-Tompkins
algorithm for preprocessing
and segmentation of ECG
data

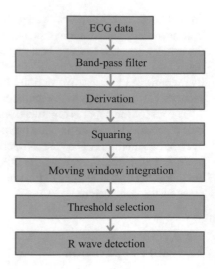

between these points are selected as the heartbeat. Next the minimum length of beats is found and other beats are trimmed in this length. In this way, the length of data is unified. In this work, the minimum length of 143 samples was found among data.

3 Feature Extraction and Classifier Selection

In this section, two feature extraction methods for classifying data are surveyed. Different types of features include: morphological features, frequency features, statistical features, linear and nonlinear features and so on. The use of each feature depends on the kind of disease which is studied. For example some diseases differ only in shape. As a result only morphological features may be adequate. On the other hand, there are also some diseases such as tachycardia in which the rate of heartbeats differ but the shape of beats do not change. In this case, the frequency features should be used too. In this study, the five classes of beats which are used (normal beats (N), right bundle branch block (RBBB), left bundle branch block (LBBB), ventricular premature contraction (VPC) and paced beat (P)) all differ morphologically. As a result, we can segment the ECG signals, beat by beat, and use only morphological features. The features which are surveyed here are taken from PCA and DWT which are described in the following.

3.1 Principal Component Analysis (PCA)

PCA, Jolliffe [3], is a method for finding the main directions of data. In other words, after segmentation of ECG records, each beat contains 143 samples. PCA looks at this data as a vector with dimension 143. So, the main distribution of the input vector which shows the most variations in data should be extracted. This direction is the first principal component of data. After finding the main direction of input vectors, the second main direction of input vectors should be calculated. PCA automatically finds the next direction of data orthogonal to former directions. This direction shows the second principal component of data. In the same way, the next main directions of variation of data are calculated which show the less important principal components of data. After finding all principal components of data, the less important directions in which the data variation is low are thrown away. As a result, the main features of data are preserved and less informative features are omitted. In the following, the implementation of PCA algorithm is described step by step.

Algorithm (1): PCA levels

(a) Data selection.
(b) Covariance matrix calculation:

$$cov(x, y) = \frac{1}{n-1} \sum_i (x_i - \bar{x})(y_i - \bar{y}) \tag{1}$$

For example if there are three axes X, Y, Z, the covariance matrix would be:

$$C = \begin{pmatrix} cov(x, x) \ cov(x, y) \ cov(x, z) \\ cov(y, x) \ cov(y, y) \ cov(y, z) \\ cov(z, x) \ cov(z, y) \ cov(z, z) \end{pmatrix} \tag{2}$$

(c) Third step: Calculation of eigenvectors and corresponding eigenvalues of covariance matrix
    If C is the covariance matrix, X_i is the eigenvector corresponding to the eigenvalue λ_i, then the following equation extracts the eigenvector X_i:

$$AX_i = \lambda_i X_i \tag{3}$$

The number of eigenvalues of covariance matrix equals the rank of covariance matrix.

(d) Forth step: Selection of components and constructing feature vector:
    In this step the notion of model reduction appears. The eigenvectors are sorted corresponding to their eigenvalues from the biggest to the smallest. Then the eigenvectors corresponding to the bigger eigenvalues are selected and the eigenvectors corresponding to the smaller eigenvalues are omitted. The resulting eigenvectors are called the feature vectors.

(e) Fifth step: reducing the size of data:
    If the data matrix is multiplied in selected eigenvectors (from right side), the
    reduced order data is derived. Final reduced order data have rows equal to
    number of data and columns equal to feature vectors. ∎

The amount of energy of signal which is saved by the feature vector depends on
the number of preserved eigenvectors. Basically, the total variance of data which is
described by each principal component equals its eigenvalue (λ_i). As a result the
overall variance which is preserved by q components, out of p components, equals
$\sum_{i=1}^{q} \lambda_i$. Therefore the percent of variance of data which is preserved by q components
out of p components equals:

$$\frac{\sum_{i=1}^{q} \lambda_i}{\sum_{i=1}^{p} \lambda_i} \tag{4}$$

The classification of data based on different number of features is surveyed in the
results section.

3.2 Discrete Wavelet Transform (DWT)

Continuous wavelet transform (CWT) was proposed as an alternative for short time
Fourier transform. Its goal was resolving the resolution problem in short time Fourier
transform. In wavelet, like short time Fourier transform, the signal is multiplied in
a function which is called wavelet here and acts as a moving window. In addition,
like short time Fourier transform, the wavelet is also separately applied on different
pieces of the signal but basically there are two main differences between wavelet and
short time Fourier transform:

- A Fourier transform is not taken from the windowed signal. This causes negative
  frequencies are not calculated here.
- In wavelet the width of window varies with the variation of frequency components.
  This is maybe the main difference of wavelet and Fourier transform.

A continuous wavelet transform is defined as:

$$CWT_x^\psi(\tau, s) = \Psi_x^\psi(\tau, s) = \frac{1}{\sqrt{|s|}} \int_{-\infty}^{+\infty} x(t)\psi^* \left(\frac{t - \tau}{s} \right) dt \tag{5}$$

where, τ, s are the translation factor and scale factor respectively. Like the concept
of scaling in the map, the large scales show an approximate view and correspond to
low frequencies whereas the small scales contain the details.

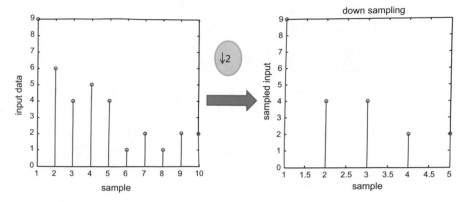

Fig. 2 Down-sampling of signal in DWT

In discrete wavelet transform (DWT), like Continuous wavelet transform, the time-scale description exists and is performed by a series of digital filters, called bank of filters. In DWT the resolution is changed by the performance of filters and the scale is changed by down-sampling or up-sampling (usually by two).

For processing a signal by DWT, the signal is passed through a low-pass filter (with impulse response $h[n]$) and a high-pass filter (with impulse response $g[n]$). These impulse responses are formulated as follows:

$$A1 = y_{low}[k] = \sum_n x[n]h[2k - n] \tag{6}$$

$$D1 = y_{high}[k] = \sum_n x[n]g[2k - n] \tag{7}$$

The relationship between low-pass filter and high-pass filter is as follows:

$$g[L - 1 - n] = (-1)^n \cdot [n] \tag{8}$$

where, L is the length of the input vector.

The output of first low-pass filter is called approximation (A1) and the output of first high-pass filter is called detail (D1). By this operation, the A1 will contain lower half frequencies of the signal and D1 will carry higher half frequencies of the input signal. The outputs A1 and D1 are the outputs of first level decomposition. The outputs of different levels are also called DWT coefficients. Due to the fact that maximum frequency of the A1 equals $\frac{\pi}{2} (rad)$, then half of the samples can be deleted. As a result, the output of each level can be down-sampled by two. This means that data can be deleted every other one. Figure 2 shows the process of down-sampling.

If the decomposition is continued after the first level on A1, then second level of decomposition is yielded. Figure 3 shows a three level decomposition of a signal. In this case, if for example the range of frequency of ECG signal is 0–100 Hz, then

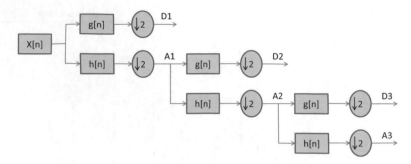

Fig. 3 Three level decomposition of a signal by DWT

A1 would have 0–50 Hz, D1 would have 50–100 Hz, A2 would have 0–25 Hz, D2 would have 25–50 Hz, A3 would have 0–12.5 Hz and D3 would have 12.5–25 Hz.

There are different mother wavelets which can be used for filtering the data. In this paper, a variety of mother wavelets are examined which include: Daubechies1or Haar, Daubechies5 (db5), Symlets2 (sym2), Symlets8 (sym8), Coiflets1 (coif1), Coiflets4 (coif4) and Biorthogonal 6.8 (bior 6.8).

3.3 DWT Feature Extraction

After decomposition of segmented ECG signals by DWT, some statistical features are extracted from the outputs of each level. In this study, the first four statistical moments of DWT coefficients, i.e. average, variance, skewness and Kurtosis of DWT coefficients are calculated as the features. The definition of mentioned moments is as follows:

- First moment (Average): $\mu = E(x) = \frac{\sum_{i=1}^{i=n} xi}{n}$
- Second moment (Variance): $V = \sigma^2 = E(x^2) - (E(x))^2$
- Third moment (Skewness): $s = \frac{E(x-\mu)^3}{\sigma^3}$
- Fourth moment (Kurtosis): $k = \frac{E(x-\mu)^4}{\sigma^4}$

3.4 Classifier Selection

After extraction of features, the features should be given to a classifier for decision making. There are a variety of features which can be used such as K-means clustering, expectation maximization (EM), C4.5 decision tree, neural networks, Bayesian network, SVM, fuzzy sets and random forest (RF). In this paper, among different classifiers, neural network has been chosen as the classifier due to its high performance, Jovic and Bogunovic [4], and simplicity. As a result, a multilayer perceptron

Table 1 Record number of utilized datasets in MIT-BIH database and the abundance of beats in total dataset

Record no.	The type of beats and their abundance in the dataset						
	Normal (N)	Right bundle branch block (R)	Left bundle branch block (L)	Ventricular prema- ture contrac- tion (VPC)	Paced beat (P)	Class 6	
						fusion of paced beat (F)	Atrial escape beat (A)
106, 109, 112, 117, 119, 121, 122, 124, 217, 219, 230, 231, 234	8874	1272	1162	480	774	8	9

once with 10 hidden neurons and later with 20 neurons in the hidden layer was chosen for classification. In addition, 90% of data was selected for training the NN and the remaining 10% was chosen for the test. Each experiment with NN was performed 50 times and the average result of experiments was inserted in the tables.

4   Simulation and Results

In this section, the results of applying the proposed machines on the ECG data from MIT_BIH database are surveyed. The dataset which is used in this study is listed in Table 1. In this dataset different classes of beats which are tagged by an annotation exist. The type of beats and the number of each type is also listed in Table 1. Due to few numbers of beats in classes (A) and (F), these categories are assumed in one class called anonymous class. The existence of such a class enables the machine to be developed later by more signals out of five realized categories.

For evaluating the performance of proposed methods, the following statistical measures are used:

- Accuracy (Acc): $Acc = \frac{TP+TN}{TP+TN+FP+FN}$
- Sensitivity (Sen): $Sen = \frac{TP}{TP+FN}$
- Specificity (Spe): $Spe = \frac{TN}{TN+FP}$
- Positive Predictive Value (PPV): $PPV = \frac{TP}{TP+FP}$

where, TP shows the number of diseases which are truly tagged as ill, TN stands for the number of safe people which are truly realized as safe, FP stands for the

Table 2 The amount of preserved variance of data and the accuracy of PCA-NN machine in classifying heartbeats

Number of features (eigenvectors)	Ratio of preserved energy $(\sum\limits_{i=1}^{q} \lambda_i \Big/ \sum\limits_{i=1}^{p} \lambda_i)$	Average accuracy after classification by NN with 10 neurons in hidden layer	Average accuracy after classification by NN with 20 neurons in hidden layer
5	0.941521985048326	91.90	94.75
10	0.998078118094588	96.70	97.65
20	0.999998756663018	98.40	98.60
50	0.999999999999790	99.15	98.85

Table 3 Details of classification of test data despite 10 features extracted from PCA and NN with 10 neurons in hidden layer

Inputs	Outputs									
	N	R	L	V	P	Other	Acc%	Sen%	Spe%	PPV%
N	672	0	0	0	0	8	98.87	98.82	98.98	99.56
R	1	87	0	0	0	6	99.28	92.55	100	100
L	0	0	104	0	0	3	99.69	97.20	100	100
V	0	0	0	24	0	16	98.36	60.00	100	100
P	0	0	0	0	74	1	99.90	98.67	100	100
Other	2	0	0	0	0	2	96.40	50	96.58	4.76

number of safe people which are falsely tagged as ill, *FN* stands for the number of ill people which are falsely realized as safe.

For the first strategy, the PCA is used for the feature extraction on the segmented data of all types. Then a neural network is applied on the selected features to classify the heartbeats. For implementing the PCA algorithm, the segmented beats with length 143 are put in rows of a matrix which yields a matrix of size $12, 562 \times 143$. Afterwards the covariance matrix is calculated as in Sect. 3.1 which yields a matrix of size 143×143. Next the eigenvalues and eigenvectors of the covariance matrix are calculated. After that, the eigenvectors are sorted according to their corresponding eigenvalues from bigger to smaller. For extracting more important features of the data, eigenvalues of the covariance matrix should be evaluated. If a prescribed value of variance of data is desirable, then according to [4], the number of required eigenvectors (q) is calculated.

In this work, the feature extraction was done with different number of eigenvectors to show the changes in quality of results despite different number of features. For this, the classification was done once with 20 features and next with 50 features. Table 2 shows the results:

As Table 2 shows, the accuracy of PCA-NN machine in classifying the heartbeats is above 90% despite even few features. Table 3 shows the detailed statistics of classification.

Table 4 The accuracy of PCA-NN machine in spite of additive white noise in classifying heartbeats

Number of features (eigenvectors)	Ratio of preserved energy $\left(\sum_{i=1}^{q} \lambda_i \Big/ \sum_{i=1}^{p} \lambda_i \right)$	Average accuracy after classification by NN with 10 neurons in hidden layer	Average accuracy after classification by NN with 20 neurons in hidden layer
5	0.941521985048326	89.55	91.85
10	0.998078118094588	93.60	90.05
20	0.999998756663018	**20.40**	**15.90**
50	0.999999999999790	**11.15**	**10.50**

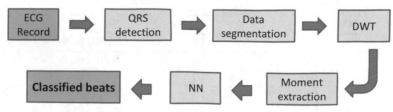

Fig. 4 The block diagram of DWT-NN machine

For evaluating the robustness of PCA-NN machine against variation of data, some white Gaussian noise (normal noise) with mean zero and standard deviation of 0.1 ($N(0, 0.1)$) is added to the segmented beats. Table 4 shows the results of former machine despite additive white Gaussian noise.

As Tables 2 and 4 show, the average percentage of accuracy changes in the Table 4 despite additional noise. In fact, by adding extra noise to each beat, the results of tests with large number of features (i.e. 20 and 50) decrease drastically. It is because the information of noise is mixed with the real information of data in the features by increment of the features. An important point from Tables 2 and 4 is that, even with few numbers of eigenvectors, good results can be achieved.

In this section the classification of heartbeats based on DWT features is declared. In this method, firstly a discrete wavelet transform is applied on the segmented signals. Then the statistical moments of the coefficients are calculated. These moments play the role of features. Finally, the features are given to the neural network for classification. Figure 4 shows this strategy for arrhythmias classification.

The classification was conducted in different scenarios i.e. with different number of moments, with different decomposition levels and with different wavelets. Table 5 shows the average percentage of accuracy despite, five decomposition levels, different number of moments and different wavelets with 10 neurons in hidden layer of NN.

As Table 5 shows the average accuracy of classification is above 92% among all classifications. Furthermore, the best accuracy is achieved with Sym2 wavelet. It is due to the similarity of this wavelet to ECG beat. Table 6 shows the statistical details of classification.

Table 5 Average accuracy of classification despite five levels of decomposition, different number of moments and different wavelets

Number of moments	Average accuracy based on applied wavelet						
	Bior6.8	Coif8	Coif1	Sym8	Sym2	Db5	Db1
2 moments	94.68	95.72	96.57	96.40	95.88	95.33	94.87
3 moments	97.14	97.99	98.23	97.56	98.36	97.26	92.60
4 moments	98.05	98.46	98.56	98.06	98.84	98.46	94.03

Table 6 Details of classification of test data despite five decomposition levels, four moments and sym2-wavelet

Inputs	Outputs									
	N	R	L	V	P	Other	Acc%	Sen%	Spe%	PPV%
N	710	0	0	0	0	4	99.60	99.44	100	100
R	0	97	0	0	0	0	100	100	100	100
L	0	0	91	0	0	0	100	100	100	100
V	0	0	0	26	0	8	99.20	76.47	100	100
P	0	0	0	0	63	1	99.90	98.44	100	100
Other	0	0	0	0	0	0	98.70	–	98.70	–

Like former machine, the robustness of DWT-NN machine against variation of the signals is evaluated by increasing a normal noise with the mean equal zero and standard deviation of 0.1 ($N(0, 0.1)$) to the segmented beats. Due to the high sensitivity of wavelet to the morphology of the signal, all results of Table 6 decreased below 50%. This result shows that the DWT-NN is more sensitive to noise than PCA-NN method.

5 Conclusions

In this paper, five classes of heartbeats are classified via two strategies based on PCA, DWT and neural network. In addition a new class, called anonymous class, is considered for rare signals which occur in heart rhythm. The first strategy used PCA and NN respectively for feature extraction and classification whereas the second strategy uses statistical features of DWT as features and NN as classifier. The ECG data which is used here is taken from MIT-BIH database. The data contains five types of ECG signals i.e. normal beats (N), right bundle branch block (RBBB), left bundle branch block (LBBB), ventricular premature contraction (VPC) and paced beat (P). The results show that both methods (PCA-NN and DWT-NN) could achieve the average accuracy more than 98% without noise. Meanwhile, PCA-NN could achieve an accuracy above 93% despite additive Gaussian noise to each beat.

Acknowledgements This work was financed by the Islamic Azad University, Nour Branch, Nour, Iran, which is gratefully acknowledged.

References

1. Alickovi E., Subasi A. (2016), Medical Decision Support System for Diagnosis of Heart Arrhythmia using DWTand Random Forests Classifier, Journal of Medical Systems, Springer, 40: 108, https://doi.org/10.1007/s10916-016-0467-8.
2. Benali R, Reguig F. B., and Hadj Selimane Z. (2012), Automatic Classification of Heartbeats Using Wavelet Neural Network, Journal of Medical Systems, Springer, 36, 883–892, https://doi.org/10.1007/s10916-010-9551-7.
3. Jolliffe I.T. (2002), Principal Component Analysis, Springer Series in Statistics.
4. Jovic A., Bogunovic N. (2011), Electrocardiogram analysis using a combination of statistical, geometric, and nonlinear heart rate variability features, Artificial Intelligence in Medicine, 51(3): 175–186.
5. Karimi Moridani M., Setarehdan S. K., Motie-Nasrabadi A., and Hajinasrollah E. (2015), Analysis of heart rate variability as a predictor of mortality in cardiovascular patients of intensive care unit, Biocybernetics and Biomedical Engineering, 35(4): 217–226.
6. Martis R. J., Acharya U. R., and Min L. C. (2013), ECG beat classification using PCA, LDA, ICA and Discrete Wavelet Transform, Biomedical Signal Processing and Control 8, 437–448.
7. Martis R. J., Acharya U. R., Mandana K. M., Ray A. K., and Chakraborty C. (2012), Application of principal component analysis to ECG signals for automated diagnosis of cardiac health, Expert Systems with Applications 39, 11792–11800.
8. MIT-BIH Arrhythmia Database Directory., Retrieved May 2, 2012, from MIT-BIH Arrhythmia Database Directory: http://www.physionet.org/physiobank/database/html/mitdbdir/mitdbdir.htm. 2001.
9. Mohseni S. S., Mohamadyari M. (2016), Heart Arrhythmias Classification via a Sequential Classifier Using Neural Network, Principal Component Analysis and Heart Rate Variation, IEEE 8th International Conference on Intelligent Systems, Sofia, Bulgaria.
10. Özbay Y. (2009), A New Approach to Detection of ECG Arrhythmias: Complex Discrete Wavelet Transform Based Complex Valued Artificial Neural Network, Journal of Medical Systems, Springer, 33:435–445, https://doi.org/10.1007/s10916-008-9205-1.
11. Pan J., Tompkins W.J. (1985), A Real-Time QRS Detection Algorithm, IEEE Transactions on Biomedical Engineering, 32(3): 230–236.
12. Poorahangaryan F., Morajab S., Kiani Sarkaleh A. (2014), Neural Network Based Method for Automatic ECG Arrhythmias Classification, Majlesi Journal of Electrical Engineering, 8(3): 33–40.
13. Sumathi S., Lilly Beaulah H., Vanithamani R. (2014), A Wavelet Transform Based Feature Extraction and Classification of Cardiac Disorder, Journal of Medical Systems, Springer, 38: 98, https://doi.org/10.1007/s10916-014-0098-x.
14. Thomas M., Kr Das M., and Ari S. (2015), Automatic ECG arrhythmia classification using dual tree complex wavelet based features, International Journal of Electronics and Communications (AEÜ), 69, 715–721.
15. Yeh Y.-C., Wang W.-J., and Chiou, C.W. (2010), A novel fuzzy c-means method for classifying heartbeat cases from ECG signals, Measurement 43:1542–1555.
16. Yu S. N., Chen Y. H. (2007), Electrocardiogram beat classification based on wavelet transformation and probabilistic neural network, Pattern Recognition Letters, 28, 1142–1150.

Printed in the United States
By Bookmasters